4632

Walley, Paul Butterfly and Moth
595.78 Wal

3510
Ss. Colman-John Neumann

P9-ASG-579

EYEWITNESS BOOKS

BUTTERFLY & MOTH

Pyralid moth,
Margaronia quadrimaculata
(China)
Smaller Wood Nymph butterfly,
Ideopsis gaura
(Indonesia)
White Satin moth caterpillar,
Leucoma salicis
(Europe & Asia)
Noctuid moth,
Diphthera hieroglyphica
(Central America)
Eyed Hawkmoth caterpillar,
Smerinthus ocellata
(Europe & Asia)
Madagascan Moon moth
Argema mittrei
(Madagascar)
Thyridid moth,
Rhondoneura limatula
(Madagascar)
Red Glider butterfly,
Cymothoe coccinata
(Africa)
Lasiocampid moth,
Gloveria gargemella
(North America)
Tailed jay butterfly,
Graphium agamemnon,
(Asia & Australia)
Jersey Tiger moth,
Euplagia quadripunctaria
(Europe & Asia)
Arctiid moth,
Composia credula
(North & South America)

Noctuid moth,
Mazuca strigitincta
(Africa)

Noctuid moth,
Apsarasa radians
(India & Indonesia)

EYEWITNESS BOOKS

BUTTERFLY & MOTH

Written by
PAUL WHALLEY

ALFRED A. KNOPF,
NEW YORK

Peacock butterfly,
Inachis io
(Europe & Asia)

Geometrid moth, *Rhodophitus simplex*
(South Africa)

Roseate Emperor moth, *Euchroa trimeni*
(South Africa)

Pyralid moth,
Ethopia roseilinea
(Southeast Asia)

Lappet moth,
Gastropacha quercifolia
(Europe & Asia)

Project editor Michele Byam
Managing art editor Jane Owen
Special photography
Colin Keates
(Natural History
Museum, London),
Kim Taylor
and Dave King

Editorial consultants
Paul Whalley and
the staff of the Natural History Museum

This is a Borzoi Book
published by Alfred A. Knopf, Inc.

First American Edition, 1988

Copyright © 1988 Dorling Kindersley Limited, London
and Editions Gallimard, Paris
Text copyright © 1988 Dorling Kindersley Limited, London
Illustration copyright © 1988
Dorling Kindersley Limited, London

All rights reserved under International and Pan-American
Copyright Conventions.
Published in the United States by Alfred A. Knopf, Inc., New York.
Distributed by Random House, Inc., New York.
Published in Great Britain by
Dorling Kindersley Limited, London.

Manufactured in Italy 0 9 8 7 6 5 4 3 2 1

Library of Congress Cataloging
in Publication Data
Whalley, Paul Ernest Sutton
Butterfly & moth / written by Paul Whalley ;
photography by Colin Keates and Dave King.
p. cm. - (Eyewitness books)
Includes index.
Summary: Photographs and text explore the behavior
and life cycles of butterflies and moths,
examining mating rituals, camouflage, habitat, growth
from pupa to larva to adult, and other aspects.
1. Butterflies - Juvenile literature.
2. Moths - Juvenile literature
[1. Butterflies. 2. Moths.] I. Keates, Colin, ill.
II. King, Dave, ill. III. Title.
IV. Title: Butterfly and moth.
QL544.2.W45 1988
595.78'022'2--dc19 88-1574

ISBN 0-394-89618-1
ISBN 0-394-99618-6 (lib. bdg.)

Color reproduction by Colourcsan, Singapore
Typeset by Windsor Graphics, Wimborne, Dorset
Printed in Italy by A Mondadori Editore, Verona

Swallowtail butterfly,
Papilio machaon
(North America,
Europe & Asia)

White Satin
moth caterpillar,
Leucoma salicis,
(Europe & Asia)

Privet
Hawkmoth
caterpillar,
Sphinx ligustri
(Europe & Asia)

African
Migrant butterfly,
Catopsilia florella
(Africa)

Cloudless Giant
Sulfur butterfly,
Phoebis sennae
(North & Central
America)

Contents

Giant Purple Emperor (Japanese national butterfly), *Sasakia charonda* (Southeast Asia)

Butterfly or moth?

BUTTERFLIES AND MOTHS are the most popular and easily recognizable of insects. Together, the two groups make up a large group (or order) of insects known as the Lepidoptera (from the Greek words for "scale" and "wing"). The Order is divided into families of butterflies and moths, containing about 165,000 known species. The division of Lepidoptera into butterflies and moths is an artificial one, based on a number of observable differences. For example, most butterflies fly by day and most moths fly by night; many butterflies are brightly colored and many moths are dull-colored; most butterflies hold their wings upright over their backs, while most moths rest with their wings flat; butterfly antennae are knobbed at the tip but moth antennae are either featherlike or plain. But despite these rules, there is not one single feature that separates all butterflies from all moths.

MEDIEVAL BUTTERFLY
A beautifully painted Red Admiral decorates a 16th-century Flemish manuscript, *Hours of Anne of Brittany.*

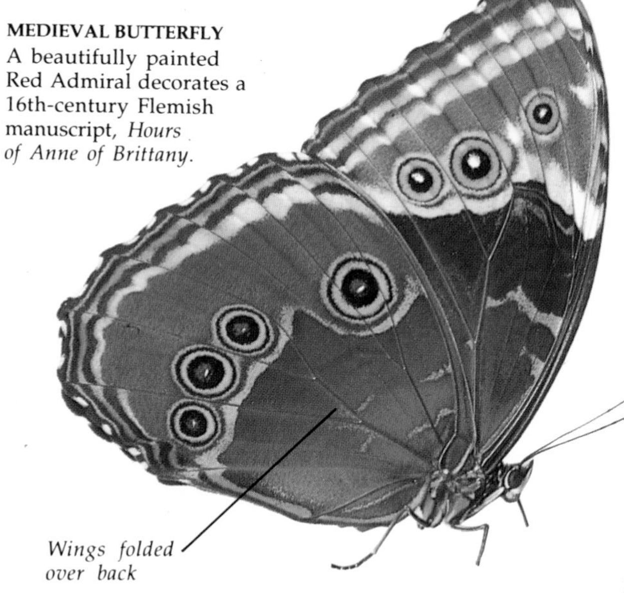

Antenna without club

Fat abdomen

SPOT THE DIFFERENCE
There are several ways to tell which of these two insects is a hawkmoth from Africa, *Euchloron megaera*, and which is a Blue Morpho butterfly, *Morpho peleides*, from Central America. Like many moths, the hawkmoth has a fat abdomen. It also has a moth's typical simple or feathery antennae, rather than the butterfly's club-tipped antennae. And if you had a magnifying glass, you could see that only the moth has a tiny hook or bristle linking its forewings and hind wings.

A short life, but a long history

It seems strange to think of graceful moths flying around giant dinosaurs, but from fossils we can tell that the first primitive moths lived about 140 million years ago. Butterflies evolved later than moths, the oldest fossils discovered so far being about 40 million years old. By the time the first humans appeared, about five million years ago, butterflies and moths were like those we see today.

AMERICAN PIONEER *left*
This 40-million-year-old specimen of a Nymphalid butterfly, *Prodryas persephone*, was found in the fossil beds of Lake Florissant, Colorado.

EGYPTIAN TOMB PAINTING
The ancient Egyptians believed that in the afterworld the dead could still hunt birds and see butterflies by the banks of the river Nile.

Lepidoptera versus the rest

After looking at the differences between butterflies and moths, it is interesting to see how they differ structurally from other orders of insects. All insects have three main divisions to their bodies: head, thorax, and abdomen. Insects have their "skeleton" around the body, not inside like mammals. If an insect's body were an undivided "tube" it would have great difficulty moving: dividing the "tube" up into segments gives greater flexibility. Structurally butterflies and moths are like all other insects; their most obvious difference is the scales covering the wings and body. Their ability to coil up the proboscis, or feeding tube, is also unique. All insects have six legs attached to the thorax, but some butterflies have shorter front legs. Insects are the only invertebrates (animals without backbones) with wings, although not all insects, including some female moths, can fly.

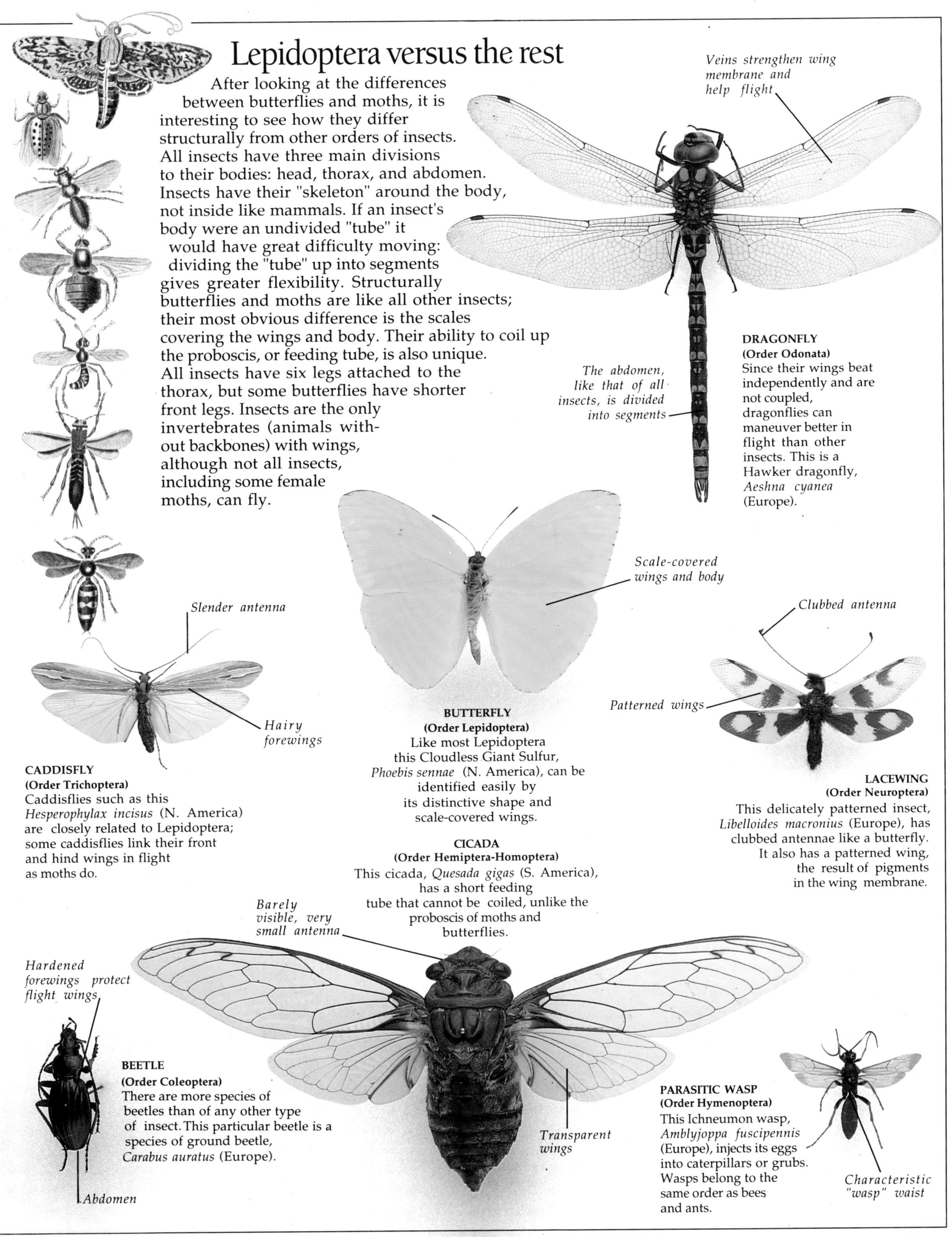

DRAGONFLY
(Order Odonata)
Since their wings beat independently and are not coupled, dragonflies can maneuver better in flight than other insects. This is a Hawker dragonfly, *Aeshna cyanea* (Europe).

CADDISFLY
(Order Trichoptera)
Caddisflies such as this *Hesperophylax incisus* (N. America) are closely related to Lepidoptera; some caddisflies link their front and hind wings in flight as moths do.

BUTTERFLY
(Order Lepidoptera)
Like most Lepidoptera this Cloudless Giant Sulfur, *Phoebis sennae* (N. America), can be identified easily by its distinctive shape and scale-covered wings.

LACEWING
(Order Neuroptera)
This delicately patterned insect, *Libelloides macronius* (Europe), has clubbed antennae like a butterfly. It also has a patterned wing, the result of pigments in the wing membrane.

CICADA
(Order Hemiptera-Homoptera)
This cicada, *Quesada gigas* (S. America), has a short feeding tube that cannot be coiled, unlike the proboscis of moths and butterflies.

BEETLE
(Order Coleoptera)
There are more species of beetles than of any other type of insect. This particular beetle is a species of ground beetle, *Carabus auratus* (Europe).

PARASITIC WASP
(Order Hymenoptera)
This Ichneumon wasp, *Amblyjoppa fuscipennis* (Europe), injects its eggs into caterpillars or grubs. Wasps belong to the same order as bees and ants.

The life of a butterfly

THE LIFE CYCLE OF A BUTTERFLY OR MOTH consists of four different stages: egg, caterpillar, pupa, and adult. The length of the life cycle, from egg to adult, varies enormously between species. It may be as little as a few weeks if the insect lives in the high temperatures of buildings where grain is stored, like some of the Pyralid moths. Other moths can live for several years. Sometimes most of the life cycle of a butterfly or moth is hidden from sight. For example, most of the life cycle of the leaf-mining moths takes place between the upper and lower surfaces of a single leaf, with only the adult going into the outside world. In a similar way, some of the wood-boring larvae of the Cossid moths may spend months, or even years, in the caterpillar stage, hidden inside a tree. Other species pass their entire life cycle much more exposed. These are usually either well camouflaged (see pp. 54-55), or distasteful to predators. There are many variations on the life cycle - some species, for example, have fewer molts in the caterpillar stage than others. These two pages illustrate the life cycle of a South American Owl butterfly, *Caligo beltrao* (also pp. 16, 23, 35).

Young caterpillar with new, green skin

Older caterpillar with brown skin is about to pupate

1 EGGS
The eggs of the Owl butterfly have delicate ribs that meet at the top. The ribbing and the structure of the shell (a tough coating like an insect's body, not a brittle one like a hen's egg) are designed to protect the egg from water loss while allowing it to "breathe" (pp. 12-13).

2 CATERPILLARS
Once the caterpillar hatches, it feeds and grows very rapidly. It molts its skin and develops a new one underneath, which stretches and allows new growth after the molt. Some species of *Caligo* are pests on bananas in Central and South America. The long, slender shape of the caterpillar helps to conceal it against the midrib of the leaves on which it feeds.

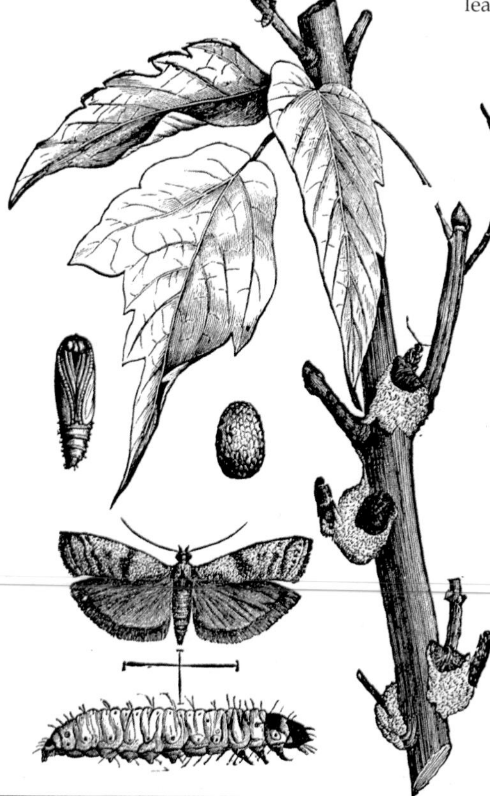

MEAT-EATING MOTH *left*
The Pyralid *Laetilia coccidivora* (N. & S. America) has a life cycle similar to other moths (pp. 36-37). It differs in the feeding habits of the caterpillar, which is predatory and eats scale insects and aphids, which it catches as it moves across the plant.

SILK SPINNER *right*
The life cycle of the Wild Silk moth, *Samia cynthia* (India), shows all the typical stages, but since it is a moth, it spins a cocoon in which to pupate (pp. 38-39). The caterpillar of this moth feeds on a variety of plants, including the castor-oil plant, (right). It also spins a dense cocoon.

WINTER SLEEPERS
The Hop Merchant or Comma, *Polygonia comma* (N. America), and the Comma, *Polygonia c-album* (Europe & Asia), are closely related Nymphalid butterflies (p. 29). Their caterpillars feed on nettles and hops, and the adult butterflies, which emerge in late summer and autumn, hibernate during winter in their adult stage (p. 51).

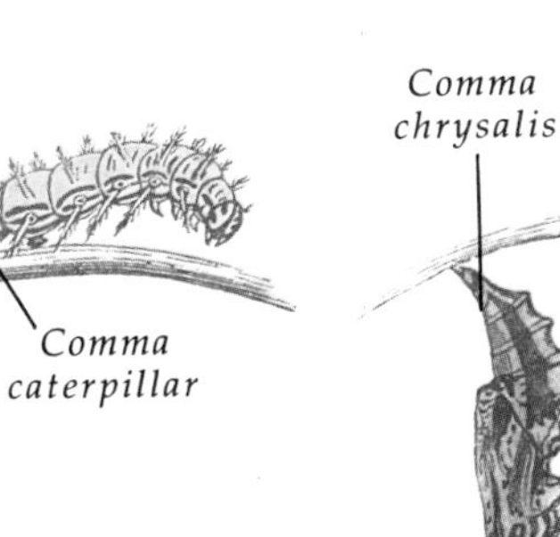

Comma caterpillar

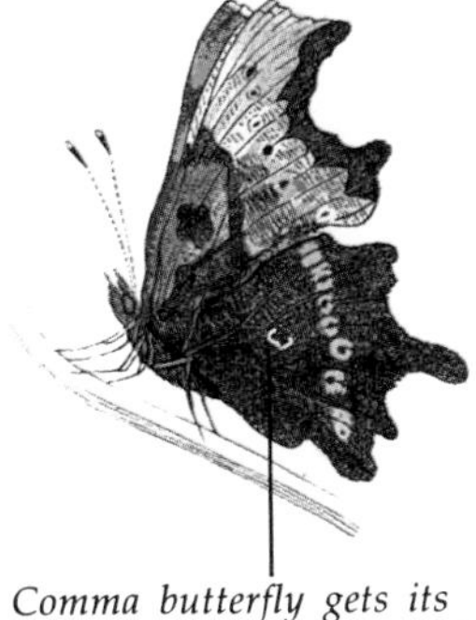

Comma chrysalis

Comma butterfly gets its name from the small white C-shape on its wing

Comma with wings open, showing its rich brown and orange patterning

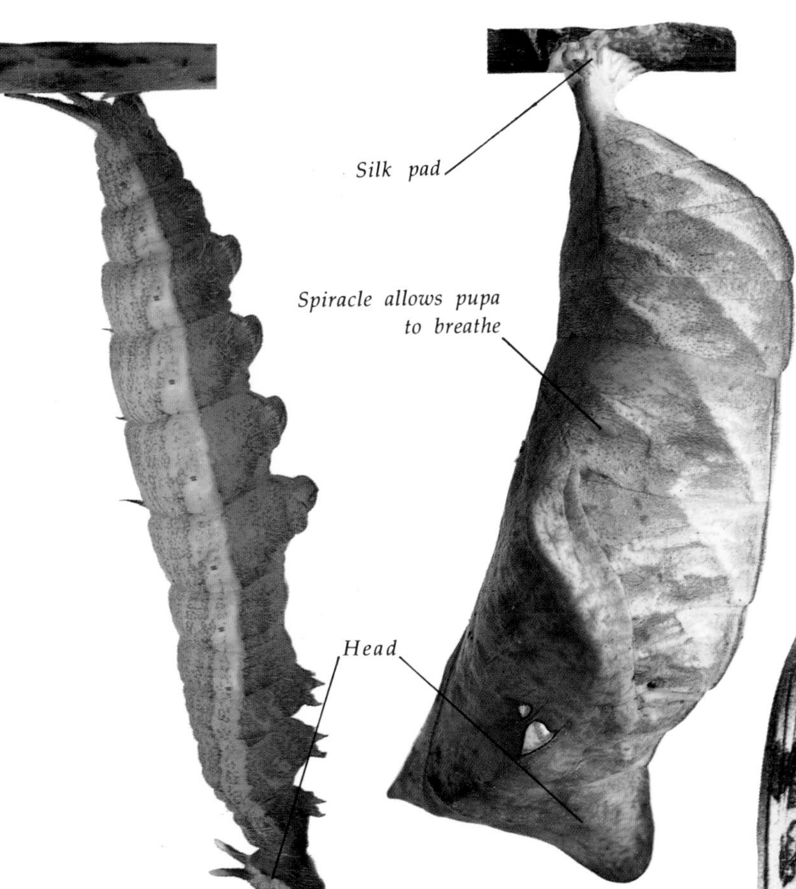

Silk pad

Spiracle allows pupa to breathe

Head

3 **PUPATING**
By this stage (pp. 20-21) the caterpillar has darkened slightly and, using the silk from the spinneret under its head, has applied a small silken pad to the plant. It attaches its hind claspers firmly to this and hangs, head down, from the stem. Underneath its skin, the skin of the next stage, the chrysalis, is forming. Gradually, with much wriggling and twisting, it will shed its caterpillar skin and shake it away - legs, head, and all - so that the completed chrysalis is revealed.

4 **CHRYSALIS**
The chrysalis (pp. 22-23), now completely formed, does not have any outside legs or antennae. Inside the chrysalis, the body of the caterpillar is broken down, special cells take over the insect's development, and gradually the adult is formed. The change from the caterpillar to the butterfly that finally emerges is one of the most remarkable events in the natural world. The oval structure on each body segment is called a spiracle and allows the chrysalis to breathe - although inactive on the outside, it needs energy for all the changes taking place inside.

5 **ADULT** *below*
The adult butterfly, so totally unlike the early feeding stages, has emerged, spread its wings , and is ready to fly (p. 35). Adult butterflies often live for only a few weeks, although a few species may survive for a year. After a time their wings often become noticeably tattered from general wear and tear; they can still fly with ragged wings, but not as well as they could before. The adult's role in the life cycle is to reproduce and scatter its eggs where they will be most likely to survive. Adult butterflies seek out new areas to live in and many can fly long distances. They usually mate as soon as possible after they emerge (pp. 10-11).

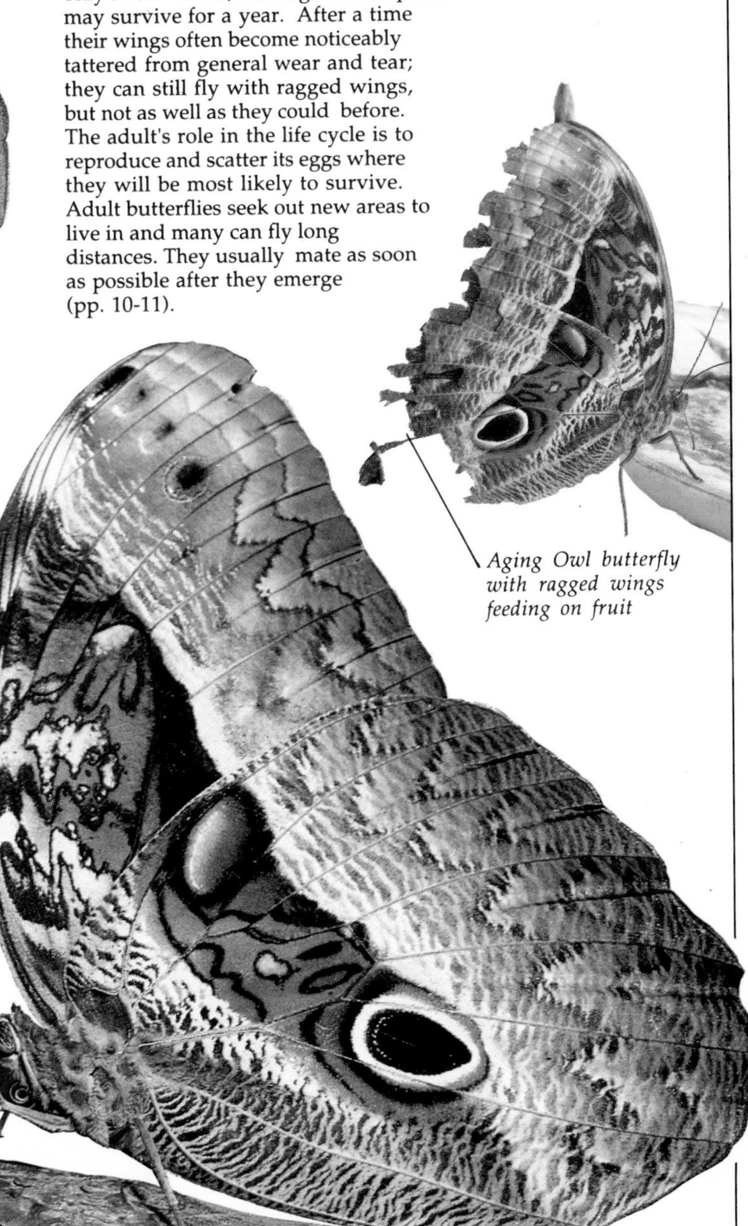

Aging Owl butterfly with ragged wings feeding on fruit

Courtship and egg laying

THE MOST IMPORTANT EVENTS in the lives of butterflies and moths are mating and the laying of eggs. The striking colors and shapes of many species are thought to attract the opposite sex; in addition, most butterflies and moths have complicated courtship behavior. As well as performing elaborate courtship flights and "dances," they often use chemicals called pheromones to attract members of the opposite sex. "Assembling" - the attraction of males to females by scent - is now known to be due to these chemicals. In butterflies it is usually the male who produces these powerful scents, while in moths it is often the female. When a male finds a female who shows an interest in him, they both land. The female holds her wings in a partly open position so that the male can land easily alongside her and continue spreading his scent. The mating pair will often tap each other with their antennae, detecting other scents which stimulate activity at close range. Mating may last for about twenty minutes, or for several hours, during which time the two insects do not move.

A 19th-century version of the butterflies' courtship dance

FROM EGG TO CATERPILLAR
This moth, *Malacosoma neustria* (Europe), has a hairy caterpillar that eats the leaves of many trees. The moth's eggs are shown on the opposite page.

BUTTERFLIES MATING
Like this pair of Sweet Oil butterflies, *Mechanitis polymnia* (S. America), most butterflies mate on a plant. They can fly while linked together, but they avoid this unless disturbed so as not to call attention to themselves. After mating, males look for another female, but the mated females look for a particular plant to lay their eggs on. Some butterflies, notably those with grass-feeding larvae, scatter their eggs, but most females actively look for a food plant for the caterpillars.

A TWO-HEADED BUTTERFLY?
A mating pair, like these two Asian swallowtails, can look like a two-headed butterfly. The tail-to-tail position links the genitalia of male and female together. The male has a complicated series of structures including claspers, which he uses to grasp the female's abdomen. The genital organs of butterflies and moths are a useful way of identifying species.

SEXUAL DIFFERENCES *right and below right*
The males and females of some butterfly species are very different in external appearance, a condition known as sexual dimorphism. An example is the Orangetip, *Anthocharis cardamines* (Europe & Asia). The males have a distinctive orange color on the wingtips, and the females have black wingtips. In some species the females are larger than the males, and a few female moths are flightless (p. 30).

Egg laying

After selecting the correct plant for the caterpillars, the female walks over a leaf, testing it carefully, presumably to make sure it belongs to the right plant species. We know that many species can detect chemicals from different plants: cabbage-eating Large and Small White butterflies, *Pieris brassicae* (Europe) and *Pieris rapae* (Europe, N. America & Australia), have been persuaded to lay eggs on plants that their caterpillars will not eat, by scientists' putting traces of extract from cabbages on the surface of the leaves.

SILK MOTH LAYING EGGS *left*
This female silk moth has laid her batch of eggs on a mulberry leaf. This moth may lay many eggs in the wild, but few of them will become adults. In artificial conditions, large numbers of moths can be raised from one egg batch (see pp. 40-41).

DELICATE OPERATION
This Pierid butterfly from Central America, *Perrhybris pyrra,* is laying its eggs on the upper surface of the leaf. She is very vulnerable to disturbance here, and a heavy rainstorm will interrupt egg laying.

SITES FOR EGG LAYING
Some species of *Heliconius* butterflies lay their eggs on tendrils of the passion-flower (above). The Lackey moth (see opposite page) lays its eggs in a ring around a twig, so that they look like part of the plant (below).

ABOUT TO HATCH *right*
These eggs of the Blue Mormon butterfly, *Papilio polymnestor* (Asia), have darkened and are about to hatch. Soon tiny caterpillars will emerge (p. 18). A Blue Mormon lays its eggs in a random pattern rather than a cluster, so there is more chance that predatory bugs will overlook some of them.

An emerging caterpillar

BUTTERFLIES AND MOTHS usually lay large numbers of eggs. The number varies greatly; some females lay over 1,000, although only a few eggs may survive to become adults. Eggs differ from one species to another in their color and in their surface texture, which can be smooth or beautifully sculptured. The two main types are a flattened oval shape, usually with a smooth surface, and a more upright shape, which often has a heavily ribbed surface. In most cases the female lays the eggs on a leaf or stem (see pp. 10-11), but some species - particularly the grass-feeding butterflies - simply release their eggs in flight. Both methods are designed to place the caterpillar as near as possible to the plant on which it feeds. These two pages show a caterpillar of a South American Owl butterfly (see pp. 8-9, 16, 23, and 25) hatching from its egg.

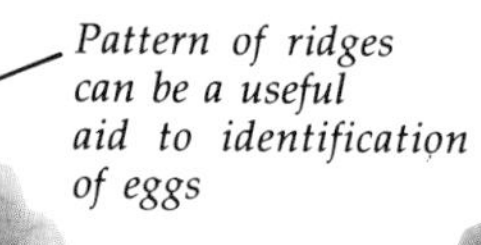

Pattern of ridges can be a useful aid to identification of eggs

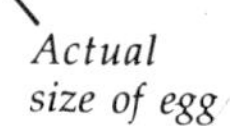

Actual size of egg

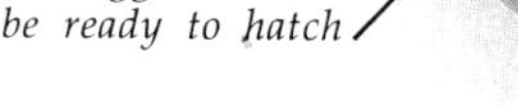

Darker color shows that egg will soon be ready to hatch

THE EGGS IN POSITION
The Owl butterfly lays its eggs in groups. The color of the individual eggs can vary in this species. They turn darker as the time of hatching gets near.

RESTING
In many temperate butterflies and moths, autumn-laid eggs usually go into a resting stage called diapause to pass the winter. This state is broken by low or fluctuating temperatures.

WARMING UP
Once winter diapause has broken, and the temperature has risen enough for the caterpillar to stand a chance of survival, the egg darkens in color as the tiny caterpillar gets ready to emerge.

CUTTING A CIRCLE
In order to hatch, the caterpillar must bite its way through the shell of the egg. This is not a hard, brittle shell like that of a hen's egg, but it still poses a tough task for the minute caterpillar: its jaws have to cut a circle big enough for the head to come out.

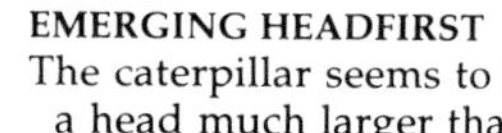

EMERGING HEADFIRST
The caterpillar seems to have jaws and a head much larger than the rest of its body, but the enormous mouthparts are useful for biting an opening in the eggshell. Nevertheless, it can be quite difficult for the small caterpillar to haul itself out of the egg headfirst. The dark spots on each side of the head are simple eyes called ocelli. The caterpillar also gets information about its surroundings from its tiny antennae.

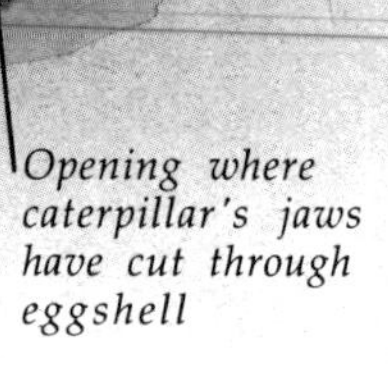

Head of caterpillar starting to appear

Opening where caterpillar's jaws have cut through eggshell

Ocelli

Antenna

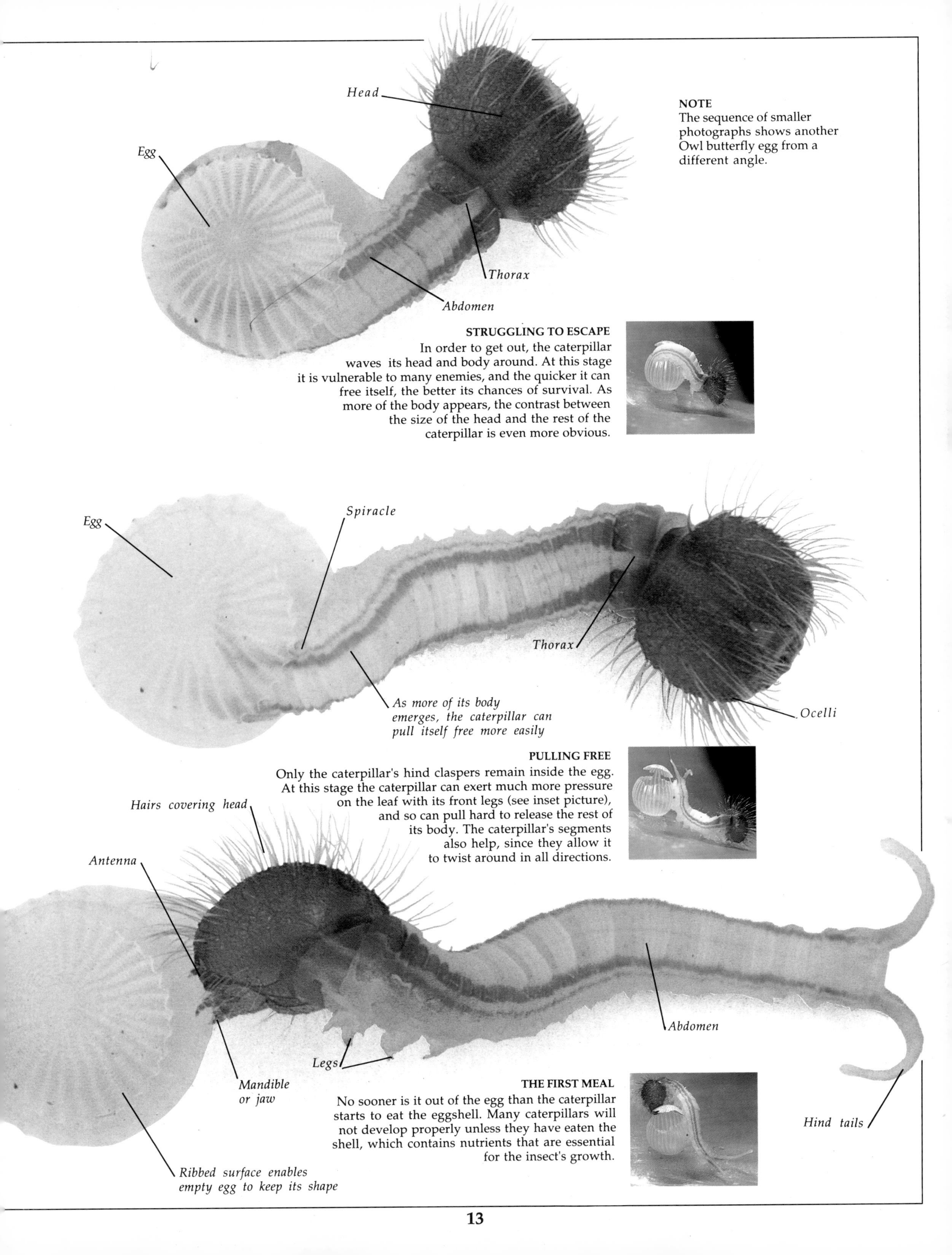

NOTE
The sequence of smaller photographs shows another Owl butterfly egg from a different angle.

STRUGGLING TO ESCAPE
In order to get out, the caterpillar waves its head and body around. At this stage it is vulnerable to many enemies, and the quicker it can free itself, the better its chances of survival. As more of the body appears, the contrast between the size of the head and the rest of the caterpillar is even more obvious.

PULLING FREE
Only the caterpillar's hind claspers remain inside the egg. At this stage the caterpillar can exert much more pressure on the leaf with its front legs (see inset picture), and so can pull hard to release the rest of its body. The caterpillar's segments also help, since they allow it to twist around in all directions.

THE FIRST MEAL
No sooner is it out of the egg than the caterpillar starts to eat the eggshell. Many caterpillars will not develop properly unless they have eaten the shell, which contains nutrients that are essential for the insect's growth.

Caterpillars

The Caterpillar talking to Alice from *Alice in Wonderland* by Lewis Carroll

IT IS A PITY THAT THE CATERPILLAR is usually dismissed simply as a "feeding tube," because it is a complex and interesting stage in the life cycle of a butterfly or moth. Caterpillars carry in their bodies the cells that eventually produce an adult insect. They molt several times during their life, discarding their outer skin to reveal a new, more elastic skin in which they can grow. Caterpillars are usually very active during this stage and need food and oxygen to grow and sustain themselves. But they do not have lungs like mammals. They take in air through small holes called spiracles in the sides of their bodies. The air passes along fine tubes, or tracheoles, from which the oxygen is extracted by the body fluid. Caterpillars have a nervous system with a primitive "brain," or cerebral ganglion, in the head. The head itself is equipped with sense organs to tell the caterpillar what is going on in the world around it. These include short antennae and often a half-circle of simple, light-sensitive "eyes," or ocelli. Also on the head are the massive jaws needed for chewing plant food. An essential feature of caterpillars, not present in the adult, is their ability to produce silk from special glands and to force it out through a spinneret under the head (pp. 40-41).

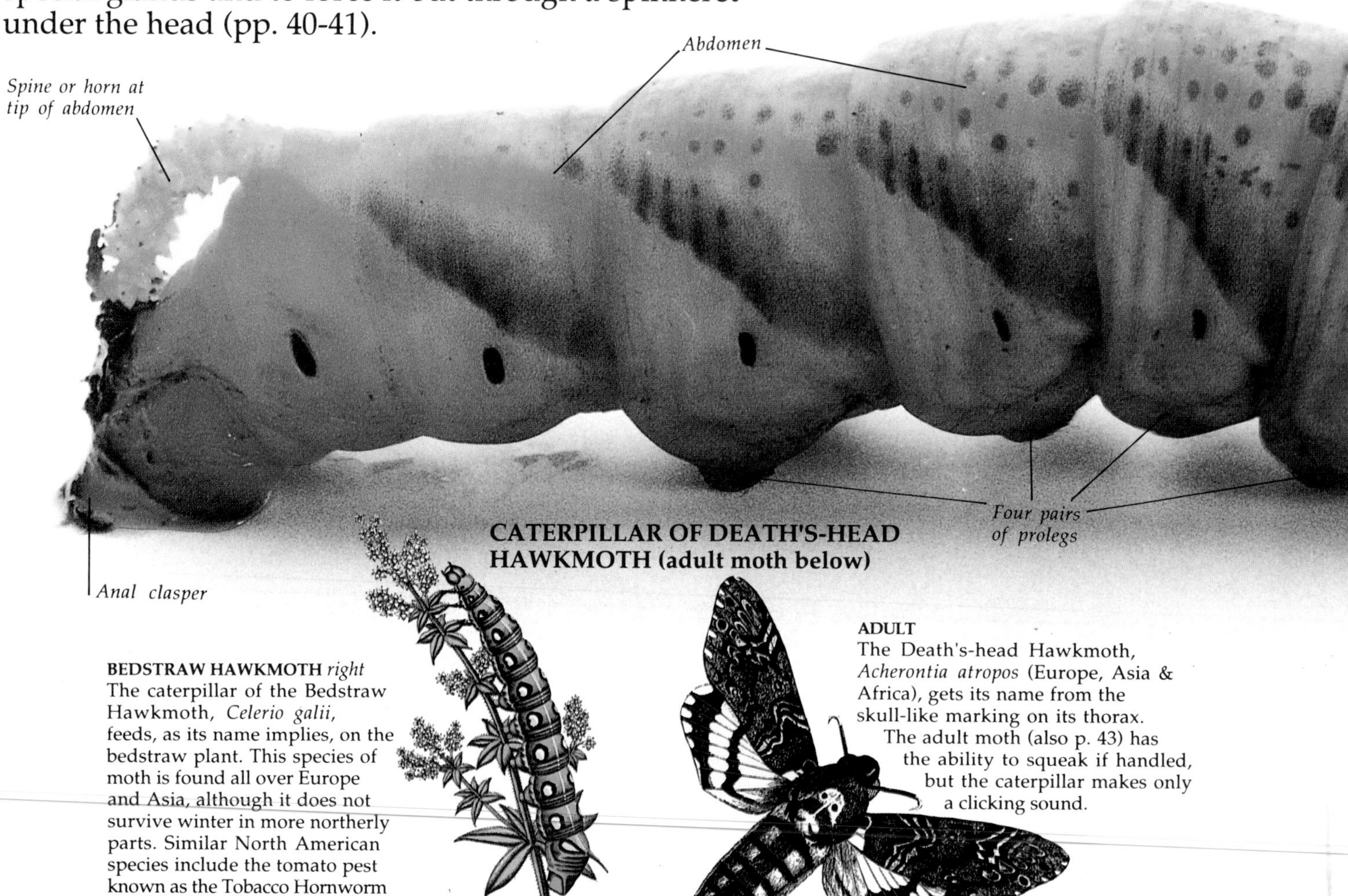

CATERPILLAR OF DEATH'S-HEAD HAWKMOTH (adult moth below)

BEDSTRAW HAWKMOTH *right*
The caterpillar of the Bedstraw Hawkmoth, *Celerio galii*, feeds, as its name implies, on the bedstraw plant. This species of moth is found all over Europe and Asia, although it does not survive winter in more northerly parts. Similar North American species include the tomato pest known as the Tobacco Hornworm or Carolina Sphinx, *Manduca sexta*.

ADULT
The Death's-head Hawkmoth, *Acherontia atropos* (Europe, Asia & Africa), gets its name from the skull-like marking on its thorax. The adult moth (also p. 43) has the ability to squeak if handled, but the caterpillar makes only a clicking sound.

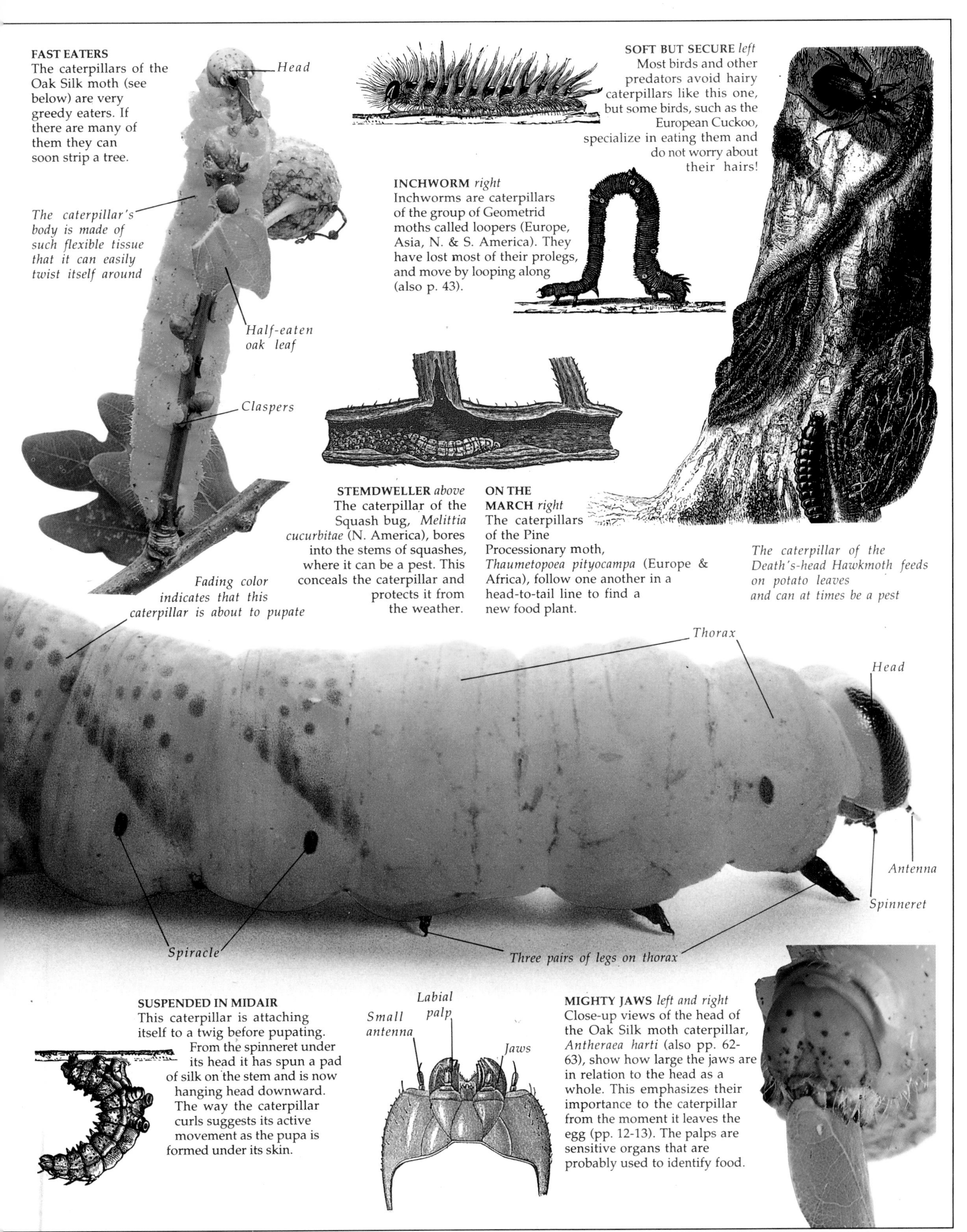

FAST EATERS
The caterpillars of the Oak Silk moth (see below) are very greedy eaters. If there are many of them they can soon strip a tree.

The caterpillar's body is made of such flexible tissue that it can easily twist itself around

Fading color indicates that this caterpillar is about to pupate

SOFT BUT SECURE *left*
Most birds and other predators avoid hairy caterpillars like this one, but some birds, such as the European Cuckoo, specialize in eating them and do not worry about their hairs!

INCHWORM *right*
Inchworms are caterpillars of the group of Geometrid moths called loopers (Europe, Asia, N. & S. America). They have lost most of their prolegs, and move by looping along (also p. 43).

STEMDWELLER *above*
The caterpillar of the Squash bug, *Melittia cucurbitae* (N. America), bores into the stems of squashes, where it can be a pest. This conceals the caterpillar and protects it from the weather.

ON THE MARCH *right*
The caterpillars of the Pine Processionary moth, *Thaumetopoea pityocampa* (Europe & Africa), follow one another in a head-to-tail line to find a new food plant.

The caterpillar of the Death's-head Hawkmoth feeds on potato leaves and can at times be a pest

SUSPENDED IN MIDAIR
This caterpillar is attaching itself to a twig before pupating. From the spinneret under its head it has spun a pad of silk on the stem and is now hanging head downward. The way the caterpillar curls suggests its active movement as the pupa is formed under its skin.

MIGHTY JAWS *left and right*
Close-up views of the head of the Oak Silk moth caterpillar, *Antheraea harti* (also pp. 62-63), show how large the jaws are in relation to the head as a whole. This emphasizes their importance to the caterpillar from the moment it leaves the egg (pp. 12-13). The palps are sensitive organs that are probably used to identify food.

Exotic caterpillars

WOOLLY BEARS *left and below*
The long hairs of many Arctiid moths - "woolly bears" - can cause allergic reactions in some people.

APART FROM BEING EFFICIENT AT FEEDING and growing, caterpillars must be able to survive in a hostile world. While caterpillars are an essential food source for birds to feed their young, it is clearly a disadvantage to be on the daily menu. This is why caterpillars have adopted a large variety of shapes and protective devices in order to survive. The caterpillars shown on the next four pages all come from tropical countries (see pp. 32-35 and 44-47), where, as in all wild places, "eat or be eaten" is very much the rule. Birds, mammals, and even certain predatory insects relish a juicy caterpillar. Fortunately for the caterpillars, many tropical species feed on plants whose contents may be poisonous. By absorbing the poisons and advertising their distastefulness with their bright colors, great numbers of caterpillars avoid an early death.

Owl Butterfly, *Caligo beltrao* (S. & C. America)

These caterpillars are not fully grown (Owl caterpillars also on p. 8)

Flambeau, *Dryas julia* (N., C. & S. America)

TINY TIGER
Like its relative the Monarch, the brightly colored caterpillar of the Plain Tiger likes to advertise its presence. It is possible that the filaments sticking out of the caterpillar's body protect it further by giving off an unpleasant smell.

Plain Tiger, *Danaus chrysippus* (Africa, S. E. Asia & Australia)

GROUP OF OWLS
The coloring of these Owl butterfly caterpillars (also pp. 8-9) makes them less noticeable along the rib of the plant. The caterpillars have a series of filaments at their heads and tails that probably help to break up their outline.

Bright stripes act as a warning to its enemies

Species of passionflower (Passiflora)

Monarch, *Danaus plexippus* (Australia, N. & S. America)

Filaments

Postman, *Heliconius melpomene* (S. America)

WARNINGLY COLORED
Monarch caterpillars can retain poisonous substances from their milkweed and dogbane food plants. Once a bird has pecked one of these caterpillars, it will usually avoid other Monarchs.

SOLITARY FEEDER

The caterpillar of the Great Eggfly, a species of butterfly found in Asia and in the Pacific region, feeds on a range of plants from cotton to some types of daisy. Adult Great Eggflies often mimic distasteful species of butterfly in order to protect themselves (see pp. 56-57 on mimicry).

DISTASTEFUL GANG

Among the most beautifully colored butterflies, Heliconiines (sometimes called Longwings) occur in the southern United States and Central and South America. Like all Heliconiines, the caterpillars of these three species feed on poisonous passionflower vines.

DANGEROUS GROUP

It is thought that the caterpillars of the Sweet Oil butterfly, *Mechanitis polymnia* (S. America), absorb poisonous substances from the leaves of the deadly nightshade plants they feed on. Although the poisons are harmless to the caterpillar and adult butterfly, they are extremely distasteful to birds and other enemies.

SWALLOWTAIL DEFENSE

The caterpillars of many swallowtail butterflies have a Y-shaped organ behind their heads. When the caterpillar is disturbed, it thrusts out two fingerlike glands, like pushing out the fingers of a glove, that emit an unpleasant smell.

Continued on next page

REARING YOUR OWN CATERPILLARS
Rearing butterflies and moths has always been a popular way of introducing children to the miracle of nature. From caterpillars collected in the wild, or from eggs obtained from the adult, the growth and development of caterpillars can be observed at close quarters (see pp. 62-63).

Adult Cracker butterfly

BABY CRACKERS
The species of *Hamadryas* variously known as Calico, Click, and Cracker butterflies are the only butterflies that make a sound as they fly. Their characteristic clicking noise is made by a special mechanism on the butterfly's wings.

Common Sailer, *Neptis hylas* (Asia)

Cracker caterpillars have black head horns and long spines

DIFFERENT COLORED CATERPILLARS
Even though they will retain their dead-leaf camouflage throughout this stage of their lives, these Common Sailer caterpillars go through a series of molts. By molting, a caterpillar not only increases its size, but also often alters its coloring and appearance.

LEAVES ON LEAVES
Although the caterpillars of the Common Sailer butterfly may seem to stand out on these individual leaves, in their natural setting their withered-leaf camouflage blends in perfectly with the surrounding foliage.

Mixed group of Asian swallowtails include: Common Mormon, *Papilio polytes*

Blue Mormon, *Papilio polymnestor;*

Great Mormon, *Papilio memnon;*

Scarlet Swallowtail, *Papilio rumanzovia*

Common Sailer, *Neptis hylas* (Asia)

SWARMING WITH SWALLOWTAILS
All the caterpillars on this plant are species of tropical *Papilio* or Swallowtail butterflies. Because most of them are early-stage larvae, it is difficult to identify individual species. The disguise taken on by this group resembles inedible bird droppings. This is obviously an extremely successful way of avoiding predatory birds.

Adult female Common Mormon butterfly

Continued from previous page

A MOTH AMONG MANY
Among the tropical caterpillars on these pages, the Silver-striped Hawkmoth is the only moth. For protection it has a black horn at one end and a fearsome look, with large yellow-ringed "eyes" on its back.

Has the characteristic horn of hawkmoth caterpillars - really a harmless long spine

Silver-striped Hawkmoth, *Hippotion celerio* (Europe, Africa, Asia & Australia)

Leopard, *Phalanta phalantha* (Africa & Asia)

Adult Silver-striped Hawkmoth

Cracker, *Hamadryas amphinome* (C. & S. America)

LEOPARDS WITHOUT SPOTS
Although this African species of butterfly doesn't look very aggressive, the popular name for it is the Leopard. Like the *Heliconius* caterpillars on pages 16 and 17, Leopards are members of the Nymphalid family, recognizable at the caterpillar stage by their spiny appearance.

Cracker, *Hamadryas feronia* (C. & S. America, sometimes Texas)

MEAL FOR A LIZARD
Although it looks as though it is about to fall victim to a hungry lizard, the caterpillar may still be able to escape if it is distasteful or spiny. It might even drop to the ground to escape the lizard.

Cracker, Guatemalan Calico, *Hamadryas guatemalena*, (C. America & sometimes Texas)

Caterpillar to pupa

THE CATERPILLAR is often regarded as simply the feeding stage in the life cycle of a butterfly, but it is a complex animal in its own right. It has to be capable of surviving in a hostile world, and it has to prepare for the vital change to the next, immobile (unmoving) stage, called the pupa, also known as the chrysalis (pp. 22-23). In moths, the chrysalis is normally contained within a cocoon (pp. 38-39). Scientists have carried out experiments to show that this remarkable change is controlled by the insect's hormones. In normal circumstances, the caterpillar must look for a place to pupate. For example, this could be a site surrounded by foliage if the insect relies on concealment for protection. If the chrysalis is protected because it is distasteful to predators, then concealment is not necessary.

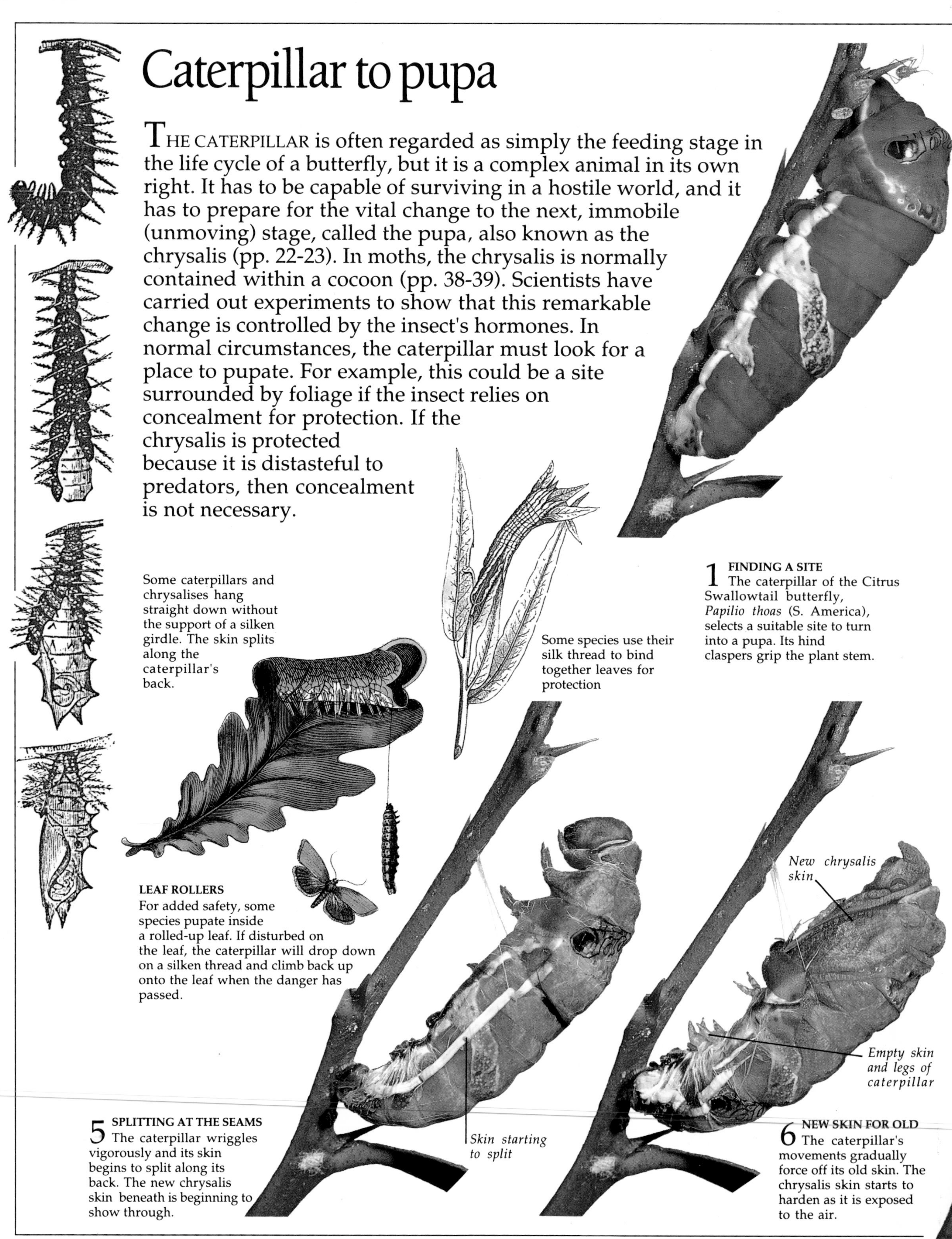

1 FINDING A SITE The caterpillar of the Citrus Swallowtail butterfly, *Papilio thoas* (S. America), selects a suitable site to turn into a pupa. Its hind claspers grip the plant stem.

Some caterpillars and chrysalises hang straight down without the support of a silken girdle. The skin splits along the caterpillar's back.

Some species use their silk thread to bind together leaves for protection

LEAF ROLLERS
For added safety, some species pupate inside a rolled-up leaf. If disturbed on the leaf, the caterpillar will drop down on a silken thread and climb back up onto the leaf when the danger has passed.

5 SPLITTING AT THE SEAMS The caterpillar wriggles vigorously and its skin begins to split along its back. The new chrysalis skin beneath is beginning to show through.

6 NEW SKIN FOR OLD The caterpillar's movements gradually force off its old skin. The chrysalis skin starts to harden as it is exposed to the air.

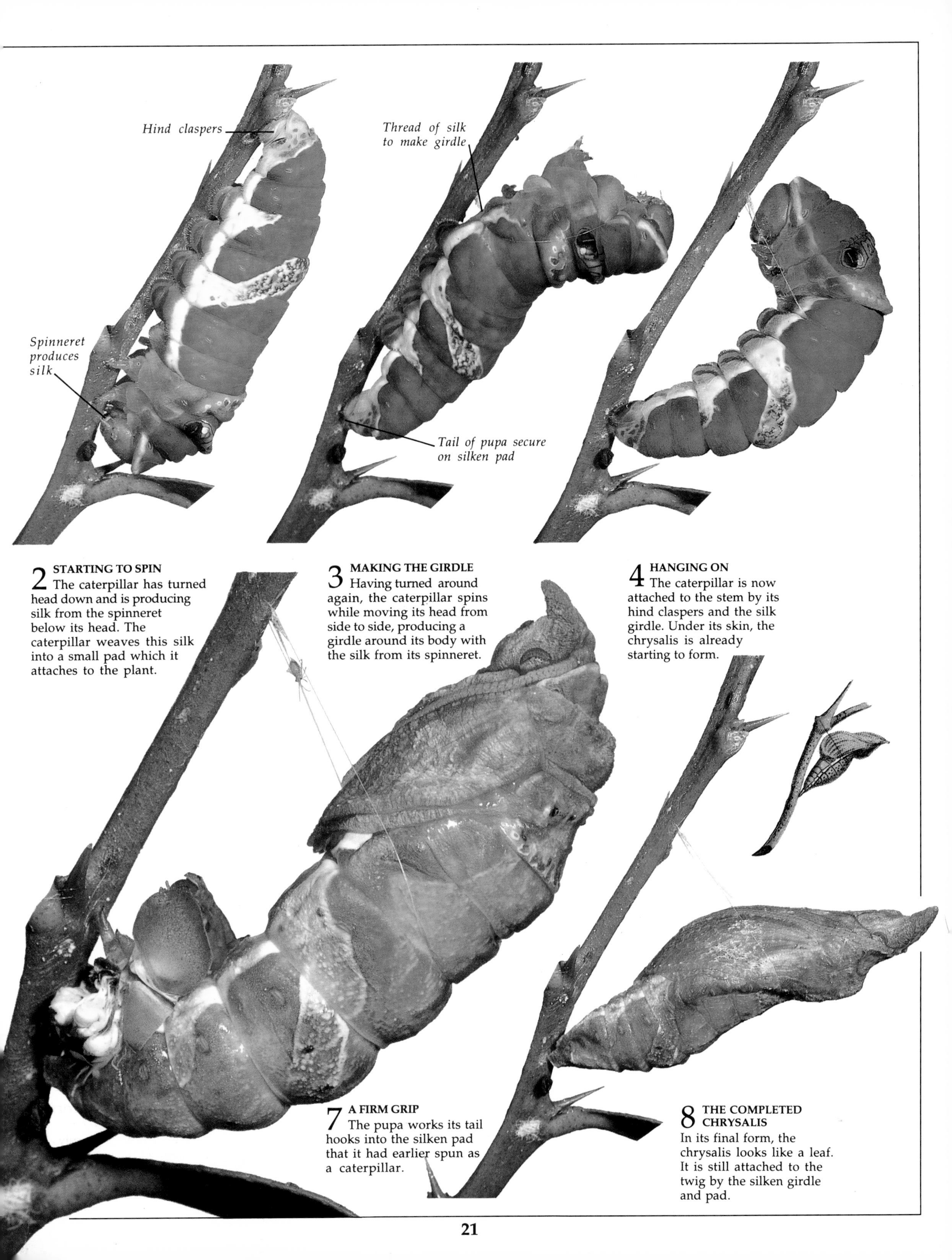

2 STARTING TO SPIN
The caterpillar has turned head down and is producing silk from the spinneret below its head. The caterpillar weaves this silk into a small pad which it attaches to the plant.

3 MAKING THE GIRDLE
Having turned around again, the caterpillar spins while moving its head from side to side, producing a girdle around its body with the silk from its spinneret.

4 HANGING ON
The caterpillar is now attached to the stem by its hind claspers and the silk girdle. Under its skin, the chrysalis is already starting to form.

7 A FIRM GRIP
The pupa works its tail hooks into the silken pad that it had earlier spun as a caterpillar.

8 THE COMPLETED CHRYSALIS
In its final form, the chrysalis looks like a leaf. It is still attached to the twig by the silken girdle and pad.

The pupa stage

THE PUPA IS THE THIRD MAJOR STAGE in a butterfly or moth's life. This is when it is transformed from a caterpillar into an adult. A butterfly pupa is usually called a chrysalis, and depending on the species and climate, it remains in this form for weeks or even months. Except for an occasional twitch, the pupa seems lifeless, but in fact amazing changes are taking place, some of which can eventually be seen through the pupal skin. Because a pupa cannot move around, the insect is far more vulnerable to predators at this time than when it is a caterpillar or an adult. For the majority of pupae their best hope of survival is to adapt their shape and color to their surroundings. The exceptions are the more brightly colored pupae which, being poisonous, are only too happy to advertise their presence. Many moths pupate underground, but few butterfly chrysalises have this added protection. Looking at the butterfly chrysalises on these pages will give some idea of how much they vary in shape and color.

THE FREAK *below*
As can be seen from these two *Calinaga buddha* (Asia) chrysalises, variation in color helps them to camouflage themselves on a wide range of backgrounds. The brown form will clearly have a protective advantage on a twig.

Wing veins

THE ARCHDUKE *left*
A close look reveals that the wing veins are visible, showing that the adult *Euthalia dirtea* (S.E. Asia) is almost ready to hatch.

MALAY LACEWING *below*
One of the important rules of camouflage is for the insect to break up its outline. The *Cethosia hypsea* (Asia) chrysalis does this by creating an irregular shape.

Shaped like dead leaf for camouflage

Spiny shape for disguise

Bright reflective gold spot distracts predators

Visible wing veins

CRUISER *left*
The resemblance to a dead and decaying leaf, and the spiny shape, helps protect *Vindula erota* (Asia) from detection by hungry predators.

THE QUEEN *above*
The chrysalis of *Danaus gilippus* (N., C. & S. America) is poisonous to predators. The poison comes from the plant on which the caterpillar feeds in its Florida Everglades habitat.

The swallowtail *Papilio machaon* (Europe, N. America & Asia) is either green or brown.

CRIMSON PATCH LONGWING *left*
As well as an irregular shape, *Heliconius erato* (S. America) has sharp spines along the wing case.

Developing wing

Developing head

Sharp spines

POSTMAN
Closely related to the Crimson Patch Longwing (left), the chrysalis of *Heliconius melpomene* (S. America) is equally well camouflaged and protective in shape.

FLAMBEAU
Dryas julia (C. & S. America) is another dark brown, rugged-looking chrysalis that gains protection from its ability to resemble woody backgrounds.

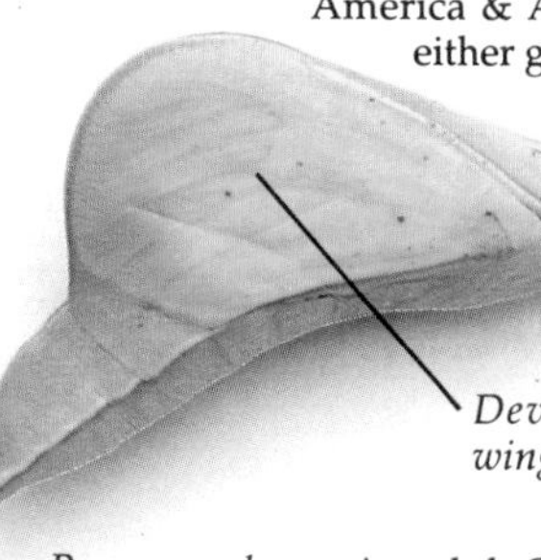

CLOUDLESS GIANT SULFUR *right*
The green, leaf-like shape of *Phoebis sennae* (N. & C. America) passes unnoticed in the vegetation of its natural habitat.

Developing wing veins

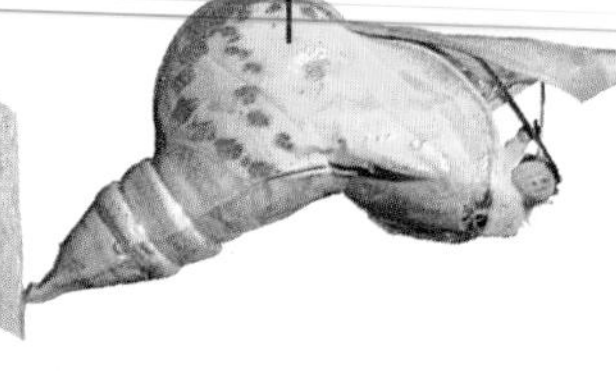

Pronounced hump in middle

An adult Cloudless Giant Sulfur (see above) beginning to break out of its chrysalis

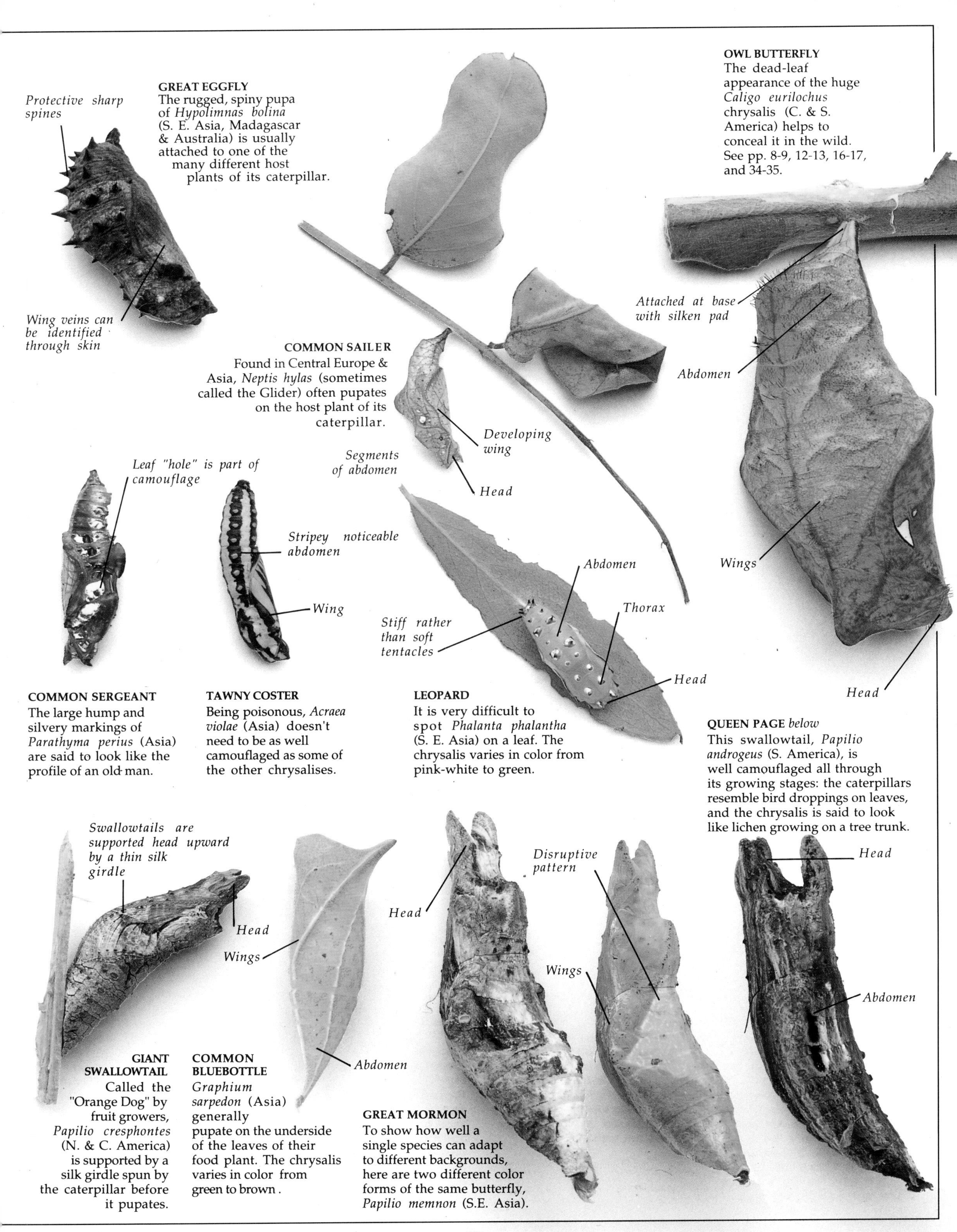

GREAT EGGFLY
The rugged, spiny pupa of *Hypolimnas bolina* (S. E. Asia, Madagascar & Australia) is usually attached to one of the many different host plants of its caterpillar.

OWL BUTTERFLY
The dead-leaf appearance of the huge *Caligo eurilochus* chrysalis (C. & S. America) helps to conceal it in the wild. See pp. 8-9, 12-13, 16-17, and 34-35.

COMMON SAILER
Found in Central Europe & Asia, *Neptis hylas* (sometimes called the Glider) often pupates on the host plant of its caterpillar.

COMMON SERGEANT
The large hump and silvery markings of *Parathyma perius* (Asia) are said to look like the profile of an old man.

TAWNY COSTER
Being poisonous, *Acraea violae* (Asia) doesn't need to be as well camouflaged as some of the other chrysalises.

LEOPARD
It is very difficult to spot *Phalanta phalantha* (S. E. Asia) on a leaf. The chrysalis varies in color from pink-white to green.

QUEEN PAGE *below*
This swallowtail, *Papilio androgeus* (S. America), is well camouflaged all through its growing stages: the caterpillars resemble bird droppings on leaves, and the chrysalis is said to look like lichen growing on a tree trunk.

GIANT SWALLOWTAIL
Called the "Orange Dog" by fruit growers, *Papilio cresphontes* (N. & C. America) is supported by a silk girdle spun by the caterpillar before it pupates.

COMMON BLUEBOTTLE
Graphium sarpedon (Asia) generally pupate on the underside of the leaves of their food plant. The chrysalis varies in color from green to brown.

GREAT MORMON
To show how well a single species can adapt to different backgrounds, here are two different color forms of the same butterfly, *Papilio memnon* (S.E. Asia).

An emerging butterfly

AS IT CHANGES FROM AN EGG TO AN ADULT a butterfly renews itself on several different occasions. When the growing stages (metamorphis) are over, all that remains is for the chrysalis to crack open and the adult butterfly to emerge. Within the unmoving chrysalis such tremendous changes have taken place that when this happens, a new creature appears to be born. The emerging butterfly shown here is a Blue Morpho, *Morpho peleides*, from Central and South America.

"The Flight into Egypt," from an illuminated manuscript, the *Hastings Hours*, c. 1480

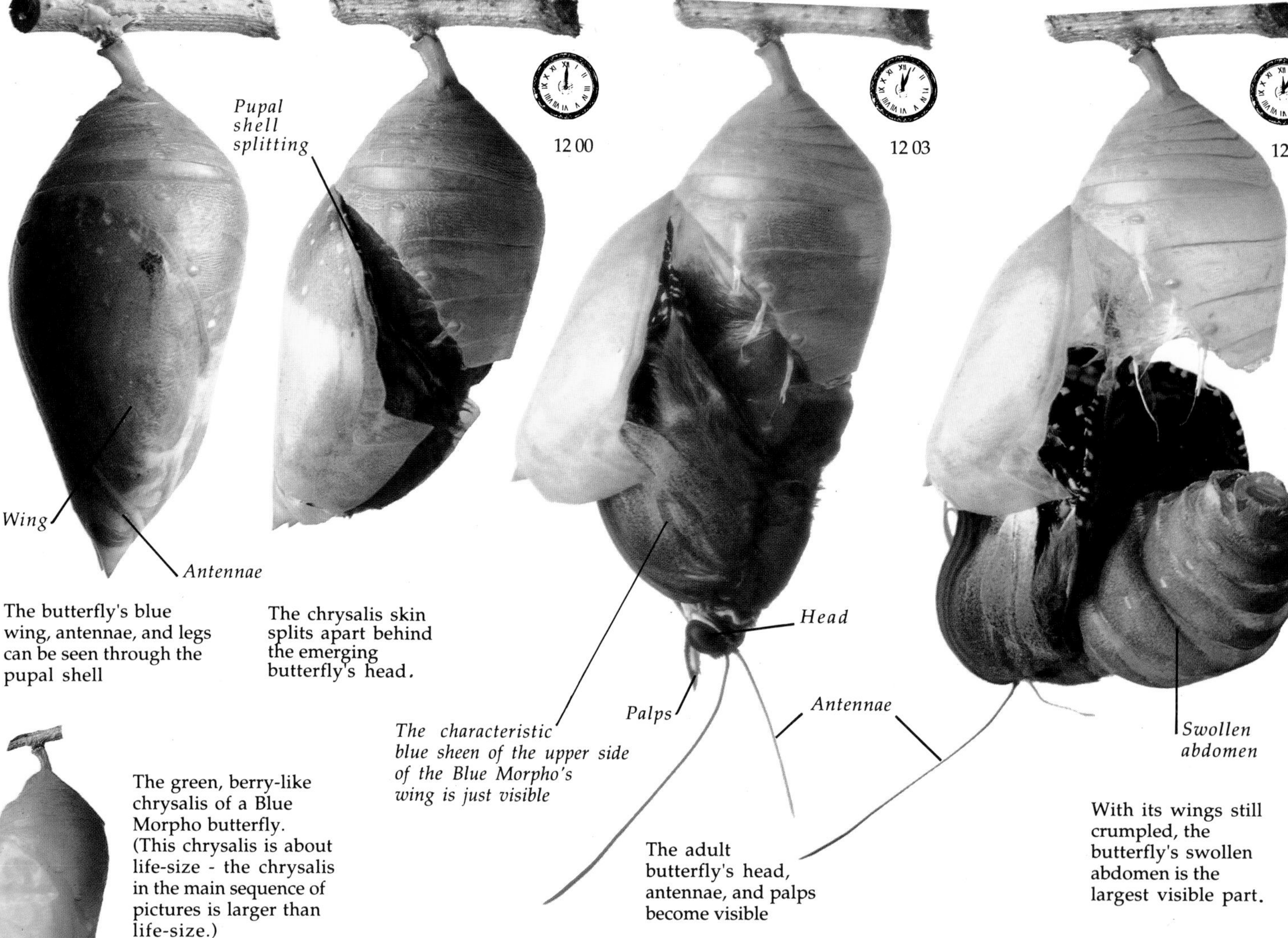

The butterfly's blue wing, antennae, and legs can be seen through the pupal shell

The chrysalis skin splits apart behind the emerging butterfly's head.

The characteristic blue sheen of the upper side of the Blue Morpho's wing is just visible

The adult butterfly's head, antennae, and palps become visible

With its wings still crumpled, the butterfly's swollen abdomen is the largest visible part.

The green, berry-like chrysalis of a Blue Morpho butterfly. (This chrysalis is about life-size - the chrysalis in the main sequence of pictures is larger than life-size.)

1 READY TO HATCH Hours before emerging, the butterfly is still developing. By now, some of the Blue Morpho's structures can be seen through the skin of the chrysalis. The dark area is the butterfly's wing, and traces of the antennae and legs are visible toward the bottom of the chrysalis. It takes about eighty-five days after the egg is laid for a Blue Morpho adult to emerge.

2 FIRST STAGE Once the insect has completed its metamorphis and is ready to emerge, it begins to pump body fluids into its head and thorax. This helps to split the chrysalis along certain weak points, so that the adult insect can begin to force its way out with its legs.

3 HEAD AND THORAX EMERGE Once the skin of the chrysalis is broken, expansion can proceed more rapidly. Inflation is due not only to the body fluids in the head and thorax, but also to the air the insect takes in. Although by now the antennae, head, and palps (sensory organs for tasting food) are visible, the wings are still too soft and crumpled for proper identification.

4 COMPLETELY FREE Having pushed its way out of the chrysalis, the butterfly's body now hangs free. At this stage, the butterfly's exoskeleton (the outside skeleton of all insects) is soft and still capable of more expansion. If, for any reason, the butterfly is damaged at this stage, or confined (perhaps by a thoughtless collector), complete expansion is not possible: all the parts harden and a crippled butterfly results.

FLY AWAY BUTTERFLY

An adult Blue Morpho butterfly, showing how the upper surface's dazzling blue sheen contrasts so vividly with the brown, spotted underside seen in the picture below (also p. 35).

2 20

5 STEADILY GROWING WINGS

With the butterfly now out of its pupal skin, the most important actions are the ejection of stored wastes from the abdomen and the expansion of the wings. As it forces blood from its body into its wings, a butterfly or moth will usually hang head- up so that the pull of gravity helps to stretch the crumpled wings.

6 BECOMING ITS FULL SIZE

By now the veins in the wings have almost filled with blood, and it is possible to see the wings visibly expanding. The expansion must take place fairly rapidly, or the wings will dry before they have reached their full size. If this happens the butterfly may be too crippled to fly.

7 WAITING TO FLY

After a period of about ten to twenty minutes, the wings reach their full size. The butterfly now waits for its wings to harden properly before it attempts to fly. Then, after an hour or so, and some preliminary opening and closing of its wings, the butterfly takes to the air. It usually flies straight to a plant or other food source for its first meal.

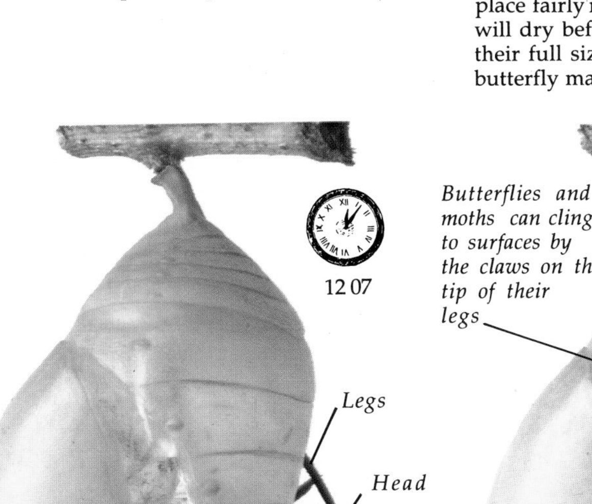

Butterflies and moths can cling to surfaces by the claws on the tip of their legs

12 12

12 20

Curled proboscis

Palps

Head

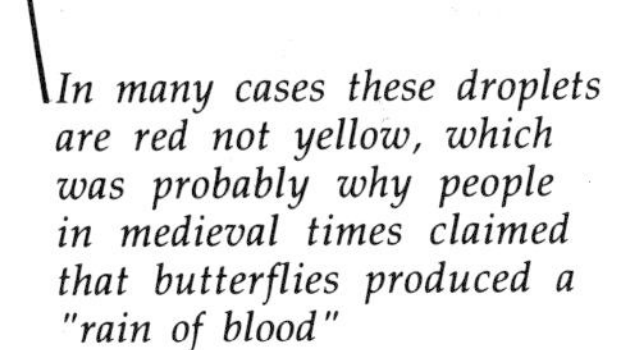

In many cases these droplets are red not yellow, which was probably why people in medieval times claimed that butterflies produced a "rain of blood"

The wing is like a bag that would expand into a balloon if it were not for tiny ligaments that hold the upper and lower membranes together

Wing veins with blood pumped into them

Once the butterfly has pushed its way clear of the chrysalis with its legs, it gets rid of waste liquid collected during the pupal stage.

The butterfly's wing patterns are now clearly visible, as are its head, palps, and proboscis

The butterfly waits with its wings held apart while they dry and harden. If it is evening, it will rest until the following day before it flies.

Butterflies

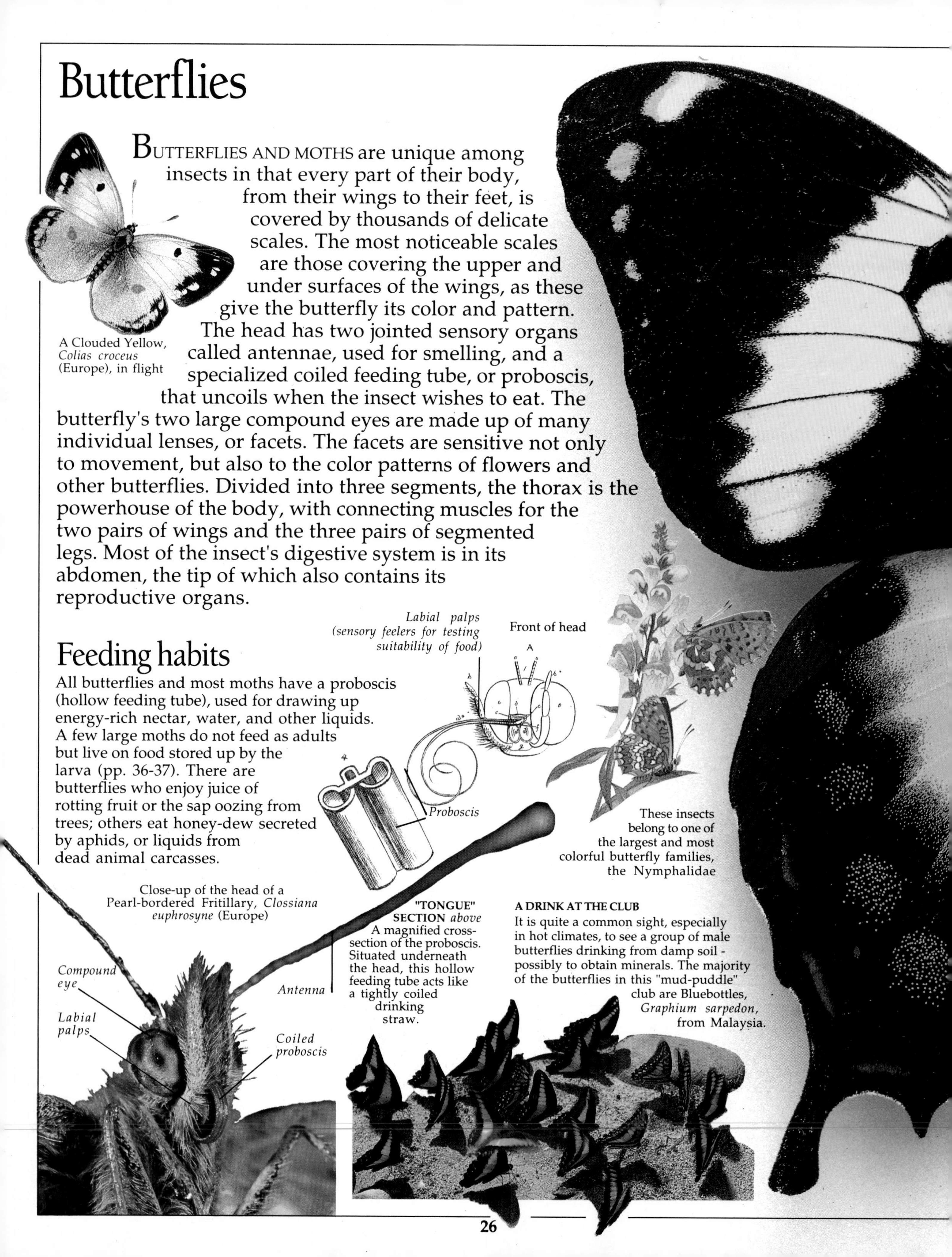

BUTTERFLIES AND MOTHS are unique among insects in that every part of their body, from their wings to their feet, is covered by thousands of delicate scales. The most noticeable scales are those covering the upper and under surfaces of the wings, as these give the butterfly its color and pattern. The head has two jointed sensory organs called antennae, used for smelling, and a specialized coiled feeding tube, or proboscis, that uncoils when the insect wishes to eat. The butterfly's two large compound eyes are made up of many individual lenses, or facets. The facets are sensitive not only to movement, but also to the color patterns of flowers and other butterflies. Divided into three segments, the thorax is the powerhouse of the body, with connecting muscles for the two pairs of wings and the three pairs of segmented legs. Most of the insect's digestive system is in its abdomen, the tip of which also contains its reproductive organs.

A Clouded Yellow, *Colias croceus* (Europe), in flight

Feeding habits

All butterflies and most moths have a proboscis (hollow feeding tube), used for drawing up energy-rich nectar, water, and other liquids. A few large moths do not feed as adults but live on food stored up by the larva (pp. 36-37). There are butterflies who enjoy juice of rotting fruit or the sap oozing from trees; others eat honey-dew secreted by aphids, or liquids from dead animal carcasses.

Labial palps (sensory feelers for testing suitability of food)

These insects belong to one of the largest and most colorful butterfly families, the Nymphalidae

Close-up of the head of a Pearl-bordered Fritillary, *Clossiana euphrosyne* (Europe)

"TONGUE" SECTION *above*
A magnified cross-section of the proboscis. Situated underneath the head, this hollow feeding tube acts like a tightly coiled drinking straw.

A DRINK AT THE CLUB
It is quite a common sight, especially in hot climates, to see a group of male butterflies drinking from damp soil - possibly to obtain minerals. The majority of the butterflies in this "mud-puddle" club are Bluebottles, *Graphium sarpedon*, from Malaysia.

HOMERUS SWALLOWTAIL, *PAPILIO HOMERUS* (JAMAICA)

RESTING POSITION *left*
In this old engraving, a Scarce Swallowtail, *Iphiclides podalirius* (Europe & Asia), is shown in a typical swallowtail resting position, with its wings folded above its body.

MAGNIFIED SCALES
A close-up view of the eyespot of a South American butterfly reveals the overlapping scales that form the wing pattern. In this picture, the tough wing veins are clearly visible.

Rows of scales form the beautiful patterns and colors of butterfly wings

WHICH FAMILY?
The veins in the wings of butterflies and moths help to keep the wing in the correct flight position. The way the veins are arranged also helps identify which family of butterfly or moth a species belongs to.

COMING IN TO LAND
With its wings slightly curved, a Peacock butterfly, *Inachis io* (Europe), is about to land on a buddleia. Butterflies have such control over their flight movements that they can make sudden landings.

Temperate butterflies

"TEMPERATE" is how we describe the regions of the earth with warm summers and cold winters.In these temperate areas butterflies are inactive during the winter months and so must be able to survive without feeding. Winter is often passed in the chrysalis stage, but there are a few butterflies in Europe and North America that pass the winter as adults, hibernating until the warmer spring weather (p. 51). The wide variety of flowers in temperate meadows and woodland clearings means that there are plenty of butterflies, although not as many as in the tropics (pp. 32-35). Temperate habitats have been increasingly destroyed and developed during recent years, and consequently butterflies are becoming less common. Their disappearance is especially sad because for most of us butterflies are the spirit of summer. Indeed, the term "butterfly" may well come from "butter-colored fly," a name for the yellow-colored Brimstone, which is one of the first European butterflies to appear each summer.

The Peacock, *Inachis io,* is one of the most common and distinctive butterflies in Europe and temperate Asia

An old engraving of a Small Copper (right) and (probably) a female Common Blue (Europe)

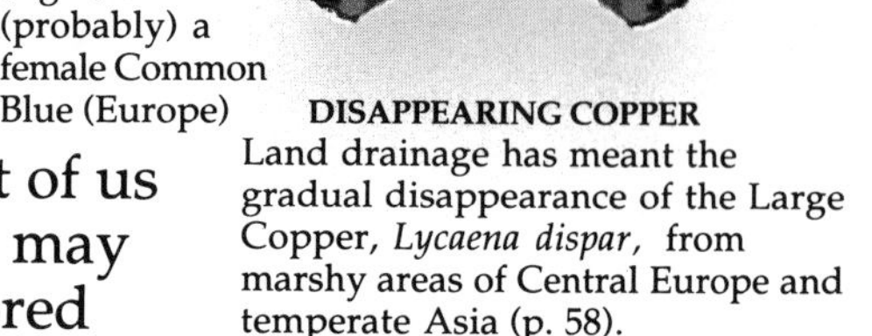

DISAPPEARING COPPER
Land drainage has meant the gradual disappearance of the Large Copper, *Lycaena dispar,* from marshy areas of Central Europe and temperate Asia (p. 58).

Grassland butterflies

BEAUTIFUL BLUE
In Europe, the Adonis Blue, *Lysandra bellargus,* is threatened in areas where its grassland habitat is under threat. It is now protected by law in France.

BENEFITING EACH OTHER
Caterpillars of the Large Blue, *Maculinea arion* (Europe), live in ants' nests, where they feed on the ant larvae. The caterpillars are not attacked by the ants, who "milk" them for a sugary solution.

GRASSLAND HABITAT
Species of butterfly whose caterpillars feed on grasses are found in meadows, shrublands, and the edges of woodlands and rivers.

BROUGHT UP ON VIOLETS
The Aphrodite, *Speyeria aphrodite,* is found in the grasslands and open woodlands of western North America. The caterpillars feed on violets.

BROWN, OR BLACK AND WHITE?
Although belonging to the Satyridae family, or "browns," the Marbled White, *Melanargia galathea* (Europe & Asia), has a black-and-white pattern.

SUCCESSFUL BROWN
The Meadow Brown, *Maniola jurtina* (Europe, Asia & Africa) is a typical well-camouflaged grassland butterfly.

This butterfly probably gets its name because it enjoys basking on walls with its wings outspread

SUN LOVER
The Wall butterfly, *Lasiommata megera* (Europe, Asia & N. Africa), is another grass-feeding species.

Somber colors on the upper and underside provide good camouflage

WIDESPREAD IN EUROPE
Although most coppers occur in Asia and America, the Purple-shot Copper, *Heodes alciphron,* is European.

A COMMON CRESCENTSPOT
The Field Crescentspot, *Phyciodes campestris,* is common in the uplands of western North America.

Woodland butterflies

Oak leaves

GREEN CAMOUFLAGE
With its brown upperside and beautiful green underside, the Green Hairstreak, *Callophrys rubi* (Europe, Asia & N. Africa), has ideal woodland camouflage.

FROM COMMA TO HOP
The Comma, also popularly known as the Hop Merchant, *Polygonia comma*, is found in a wide range of woodlands in North America and Europe. It belongs to a group of butterflies called anglewings, in which different species have been named after their distinctive wing markings.

Irregular dead-leaf outline and pattern gives effective camouflage

Upperside of Comma

The name Comma refers to a mark on the underside of the hind wing

Underside of Comma

MIXED WOODLAND HABITAT
Because of the variety of food sources, more species of butterfly are found in mixed woodland than in any other habitat. Some species of butterfly can be found flying at a low level in shady woodland clearings, and others live high among the treetops. Other species of butterfly live along woodland edges and in areas where people have cleared forests.

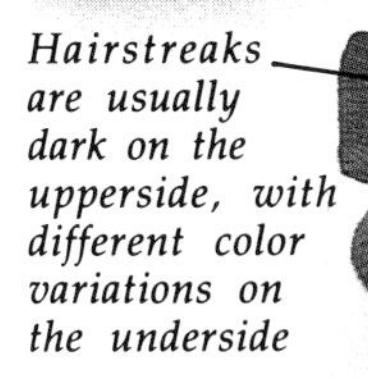

Hairstreaks are usually dark on the upperside, with different color variations on the underside

FROM STREAMS TO CANYONS
The Acadian Hairstreak, *Satyrium acadica* (N. America), occurs in damp meadows, by streams, and in canyons.

OAK FOREST RESIDENT
The Purple Hairstreak, *Quercusia quercus*, is one of a number of European and Asian species of hairstreaks.

Only males have shimmering wing scales that reflect purple when the light is at a particular angle

FEEDS ON DEAD ANIMALS
Although the Purple Emperor, *Apatura iris* (Europe & Asia), flies high up in trees, the males are attracted to the ground to feed on rotting animal carcasses.

BROKEN PATTERN
The color pattern of the Common Glider, *Neptis sappho* (Europe & Asia), is less noticeable in the dappled light of a woodland glade.

This is the form from southern Europe - the Speckled Wood in northern Europe has creamy-white markings

WOODLAND CAMOUFLAGE
The color pattern of the Speckled Wood, *Pararge aegeria* (Europe, Asia & N. Africa), makes it especially difficult to spot in patches of sunlight.

INTO THE WOOD
The Woodland Grayling, *Hipparchia fagi* (Europe & Asia), blends with bark patterns on tree trunks.

Pine White caterpillars sometimes completely strip pine trees of their leaves

PINE PEST
The adult Pine White, *Neophasia menapia* (N. America), lives among the pine trees on which its caterpillars feed.

FLYING TORTOISES
Large Tortoiseshells, *Nymphalis polychloros* (Europe & Asia), often occur in wooded uplands.

Mountain butterflies

OF ALL THE ENVIRONMENTS in which butterflies and moths live, the short summers, cold nights, and strong winds of the mountains and the treeless Arctic tundra are surely the most hostile. Insects have to adapt to harsh climates, which is why many mountain butterflies are darker than related species from lowland areas. Because darker colors absorb sunlight more easily, the insects can warm up rapidly in the early morning, when the air temperature is low. Other mountain and Arctic butterflies retain heat through the long, hairy scales that cover their bodies. In the rocky terrain of high mountains, many species lay their eggs in rocky crevices rather than on plants, and the short summer season means that they can breed only once a year. Butterflies living in constant strong winds fly in low, short bursts to avoid being blown away, and many flatten themselves against rocks when at rest. Although few species are found at very high altitudes, there are some notable exceptions of butterflies living on the very edge of the snow line in mountain ranges such as the Himalayas.

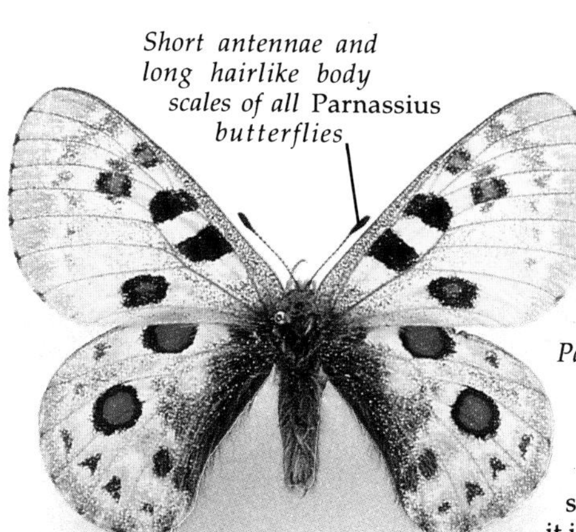

Short antennae and long hairlike body scales of all Parnassius *butterflies*

Small Apollo, *Parnassius phoebus* (Europe, Asia & N. America)

HIGH FLYER
The beautiful Apollo, *Parnassius apollo*, is found on some of the higher mountains of Europe and Asia. Because its many local forms are much sought after by collectors, it is now protected by law in most of Europe.

Male

Female

Although not a mountain species, the female Mottled Umber, Erannis defoliara, *is wingless*

FLIGHTLESS MOTHER
Some female moths are wingless, which can be an advantage on mountains where a moth could be blown away while laying its eggs.

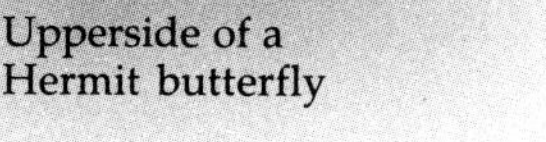

Upperside of a Hermit butterfly

Underside of a Hermit butterfly

STONY SURVIVOR
One of the best ways for a butterfly to survive in a bare, rocky environment is to be well camouflaged at all times. The Hermit butterfly, *Chazara briseis*, can be found on dry stony slopes in Central and southern Europe, and the Middle East.

HIGH MOUNTAIN HABITAT
Butterflies and moths are attracted to alpine meadows by the numerous summer flowers. This scene could be in the American Rockies, the European Alps, or the Asian Himalayas.

A species of Asian Pontia *has been found at 14,000 ft (4,250 m) in northern India*

CLOSE TO THE SNOW LINE
The Peak White butterfly, *Pontia callidice* (Europe & Asia), a relation of the Western White, *Pontia occidentalis* (N. America), is found near the snow line on high alpine mountains.

FRIEND OR ENEMY?
Found on higher ground in mainland Europe and Asia, the caterpillar of the Idas Blue, *Lycaeides idas*, spends the winter in ants' nests. The higher the altitude, the smaller this tiny butterfly becomes.

HIGH OR LOW
While Zephyr Blue, *Plebejus pylaon*, colonies are found in a variety of grassy habitats in Europe and Asia, the subspecies *trappi* occurs only in the central and southern European highlands.

MARSHES TO MOUNTAINS
During the short Arctic summer, the Palaeno Sulfur (or Moorland Clouded Yellow), *Colis palaeno,* occurs in North America, Europe, and Asia in marshy and mountain areas.

ROCKY MOUNTAIN VISITOR
The Piedmont Ringlet, *Erebia meolans,* is often seen on the rocky slopes of southern Europe. A number of related species of *Erebia* occur in the mountains of North America.

NOT SCARCE BUT NOT THERE
Once believed to occur in Britain, the Scarce Copper, *Heodes virgaureae,* in fact only occurs in the mountainous areas of Central Europe.

WAY OUT WEST
Although the Northern Marblewing, *Euchloe creusa,* comes from the mountains of the American West, it has several European relatives. Its name comes from the pattern on its hind wings.

DIFFICULT TO FIND
Cynthia's Fritillary, *Euphydryas cynthia,* occurs only in the European Alps and the mountains of Bulgaria. There are many species of fritillary, both in North America and Europe.

MOUNTAIN FLOWERS
Butterflies and moths flourish in the wild alpine pastures, where few people go. Heather (below) is typical of the plants that attract mountain and tundra butterflies in high summer.

MOUNTAIN BEAUTY
The striking-looking Bhutan Glory, *Bhutanitis lidderdalei,* comes from the mountain forests of Thailand and India. In Thailand, many of these butterflies are killed and exported to collectors.

Prominent tails on hind wings distract birds from pecking at more vulnerable parts of the body

Large eyespots can startle predatory birds

ONE OF A KIND
Butler's Mountain White, *Baltia butleri,* comes from the Himalayas. There are similar species of mountain whites in South America.

Exotic butterflies

NO REGION has so many marvelously colored and patterned butterflies as the tropics - the hot areas of the earth that are near the Equator. The range of color and pattern is quite remarkable, although we can only guess why some of these butterflies are so brightly colored: it may be for display to attract a mate, but it also may be a form of camouflage. In the bright tropical forest, with its deep shadows and intense patches of sunlight, a brightly colored butterfly may not stand out. In many species the bright colors warn predators that the butterflies are distasteful. But however wide a range of color and pattern tropical butterflies display, their shapes do not vary as much as moths' do. There are also far fewer species of butterflies than moths in the world.

Trailing tails typical of many swallowtails

FOREST MYSTERY *above*
Because it lives high up in the dense, steamy forests of New Guinea, it is hardly surprising that little is known about the Purple Spotted Swallowtail, *Graphium weiskei.*

Unusually-shaped wings measure 5 in (127 mm) across

BRILLIANT GREEN BEAUTY
Some of the most lovely butterflies are the large Birdwing Swallowtails of the New Guinea region. Species such as the *Ornithoptera priamus* are protected against overselling, but not from the destruction of their habitat.

Males of priamus *group of Birdwings have golden fringes on hind wings for transferring scent during courtship*

TROPICAL RAIN FOREST HABITAT
Although tropical rain forests can be found in Southeast Asia, northeast Australia, the South Pacific islands, and Central Africa, the place to look for the widest range of tropical butterflies is Central and South America. In these dense areas, with no winter, abundant rainfall, and a huge variety of plants, butterflies have the perfect living quarters.

WET OR DRY *below*
When it flies, the highly visible Mother of Pearl, *Protogoniomorpha parhassus* (Africa), catches the light. But once at rest in the rain forest, its color and shape make it look like a dead leaf. Mother of Pearls have wet- and dry-season forms: in the wet season the butterflies are smaller than in the dry season.

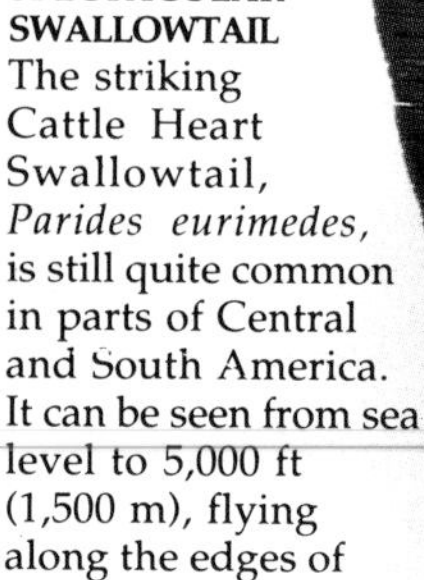

SPECTACULAR SWALLOWTAIL
The striking Cattle Heart Swallowtail, *Parides eurimedes,* is still quite common in parts of Central and South America. It can be seen from sea level to 5,000 ft (1,500 m), flying along the edges of rain forests.

Tail-less swallowtail

Irregularly shaped wing

CAMOUFLAGE BAND
Very little is known about the distinctive-looking *Taenaris schoenbergi* (New Guinea). It is likely that the band across its forewings is a form of camouflage, concealing the true outline of the butterfly when it is resting.

The waxy green leaves and brilliant red flowers of a hanging *Columnia* are typical of tropical plants (right)

DISTRACTING TAILS *below*
Species of swallowtails with distinctive trailing tails are known as "clubtails." The Yellow-bodied Clubtail, *Atrophaneura neptunus,* lives in the humid forests of Malaysia.

BUTTERFLY FEEDING
A 19th-century painting by Marianne North of an exotic-looking butterfly feeding on nutmeg juices in Jamaica.

BRIGHT TEMPTRESS
Jezebel is the popular name for several species of *Delias* butterflies. The *Delias belisama* flies in the mountainous areas of Indonesia, where its beautiful bright orange color can easily be spotted in the damp, misty tropical forests.

KEEPING OUT OF SIGHT *right*
Little is known about the rare *Dynastor napoleon* except that it probably flies only at dusk in the rain forests of Brazil. As the underside of the huge wing is patterned like a leaf, the butterfly is well camouflaged at rest.

ROYAL ROADRUNNER
If you could find the rare Royal Assyrian, *Terinos clarissa,* it would be flying at low altitudes near roadsides, quarries, and rocky outcrops in Malaysia and Indonesia. The caterpillar is covered with long, branched spines.

LIZARD ENEMY
Tropical butterflies need defense mechanisms such as trailing tails, frightening eyespots, and poisonous scales to escape from predators such as lizards.

Continued on next page

MORE COLORFUL MALE A collector brought this Nymphalid butterfly, *Myscelia orsis,* back from Paraguay many years ago. The vivid blue of this male contrasts sharply with the duller and more strongly patterned female.

NOT REALLY A MIMIC *right* The male Danaid Eggfly looks quite different from the female (below). Other popular names for this species, found in Africa, N. & C. America, India & Australia, are the Mimic, Diadem, or Six-continents.

PART OF A GROUP *below* The Theclid Hairstreak, *Amblypodia morphina,* is one of a number of similar Southeast Asian butterflies.

PERFECT LOOKALIKE *above* The nonpoisonous female Danaid Eggfly, *Hypolimnas misippus,* is a brilliant impersonator of the poisonous Tiger butterfly, *Danaus chrysippus* (mimicry pp. 56-57).

BRIGHT UNDERWINGS *right* Like many butterflies, the underside (illustrated) of the Malay Lacewing, *Cethosia hypsea,* has a more interesting pattern than the upperside. This specimen was collected in Borneo.

POISONOUS GIANT A wingspan of up to 10 in (250 mm) makes the African Giant Swallowtail, *Papilio antimachus,* the largest African butterfly. The butterfly is believed to be extremely poisonous and is avoided by its enemies in the rain forest.

An engraving of a tropical Swallowtail, *Papilio crino* (Sri Lanka)

BEAT THIS CAMOUFLAGE *below* Few butterflies have a more interesting resting camouflage than the South American butterfly *Coenophlebia archidona.*

BANANA EATER *left* The adults of *Taenaris macrops* (New Guinea) love feeding on ripe bananas. The caterpillars of some *Taenaris* butterflies feed on banana leaves.

Continued from previous page

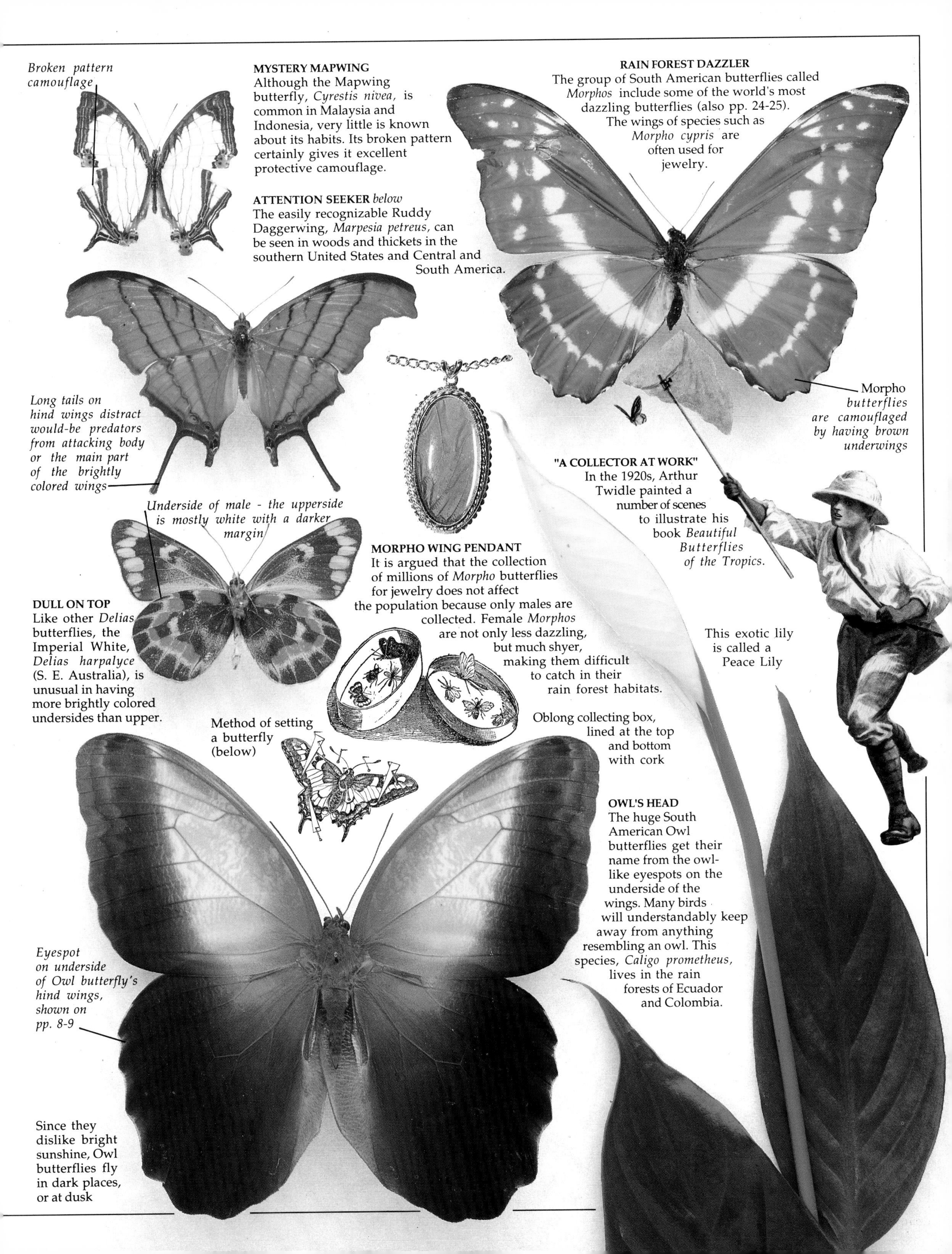

Broken pattern camouflage

MYSTERY MAPWING
Although the Mapwing butterfly, *Cyrestis nivea,* is common in Malaysia and Indonesia, very little is known about its habits. Its broken pattern certainly gives it excellent protective camouflage.

ATTENTION SEEKER *below*
The easily recognizable Ruddy Daggerwing, *Marpesia petreus,* can be seen in woods and thickets in the southern United States and Central and South America.

Long tails on hind wings distract would-be predators from attacking body or the main part of the brightly colored wings

RAIN FOREST DAZZLER
The group of South American butterflies called *Morphos* include some of the world's most dazzling butterflies (also pp. 24-25). The wings of species such as *Morpho cypris* are often used for jewelry.

Morpho *butterflies are camouflaged by having brown underwings*

"A COLLECTOR AT WORK"
In the 1920s, Arthur Twidle painted a number of scenes to illustrate his book *Beautiful Butterflies of the Tropics.*

Underside of male - the upperside is mostly white with a darker margin

DULL ON TOP
Like other *Delias* butterflies, the Imperial White, *Delias harpalyce* (S. E. Australia), is unusual in having more brightly colored undersides than upper.

MORPHO WING PENDANT
It is argued that the collection of millions of *Morpho* butterflies for jewelry does not affect the population because only males are collected. Female *Morphos* are not only less dazzling, but much shyer, making them difficult to catch in their rain forest habitats.

This exotic lily is called a Peace Lily

Method of setting a butterfly (below)

Oblong collecting box, lined at the top and bottom with cork

OWL'S HEAD
The huge South American Owl butterflies get their name from the owl-like eyespots on the underside of the wings. Many birds will understandably keep away from anything resembling an owl. This species, *Caligo prometheus,* lives in the rain forests of Ecuador and Colombia.

Eyespot on underside of Owl butterfly's hind wings, shown on pp. 8-9

Since they dislike bright sunshine, Owl butterflies fly in dark places, or at dusk

Moths

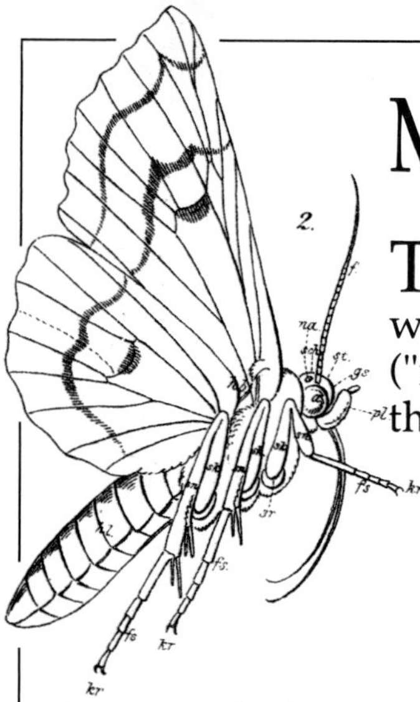

THERE ARE AT LEAST 150,000 DIFFERENT SPECIES of moths compared with some 15,000 butterflies. *Nachtschmetterlinge* ("night butterflies"), the German word for moths, clearly reflects the popular view of their behavior. While it is true that the majority of moths fly at dusk or during the night, quite a large number are day-flyers (pp. 48-49). Although moths such as the silkworm (pp. 40-41) are helpful to people, a few species of moth are harmful. These include the moths that destroy crops, fruit, and trees; the clothes moths that damage woolen goods; and moths that spread diseases in cattle by feeding on the moisture around their eyes (p. 56). The majority of moths are harmless, pollinating flowers and forming a vital part of the complex web of life.

An engraving showing the main parts of a moth, with the darker lines representing the fascia, which are part of the wing pattern

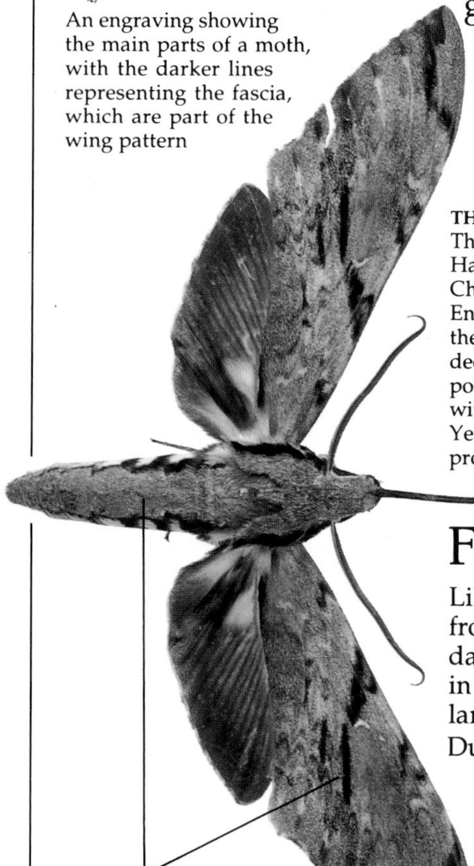

THE LONGEST TONGUE?
This amazing proboscis belongs to the Darwin's Hawkmoth, *Xanthopan morganii*, from Madagascar. Charles Darwin, the celebrated 19th-century English naturalist, knew of an orchid in which the nectar was at its base some 12 in (30 cm) deep. As the orchid obviously needed to be pollinated, Darwin thought there must be a moth with a proboscis between 12 and 13 in (30 to 35 cm). Years later, the discovery of this hawkmoth proved that Darwin's theory was correct.

Characteristic thick body and long forewings of all hawkmoths. All the moths in this group are powerful flyers.

Feeding

Like butterflies, most moths take nectar from flowers. You may be able to see day-flying moths (pp. 48-49) hovering in front of a flower as they feed. Many large moths do not feed at all as adults. During its short adult life, the Indian Moon moth (right and below) lives entirely off food stored in its body during the caterpillar stage.

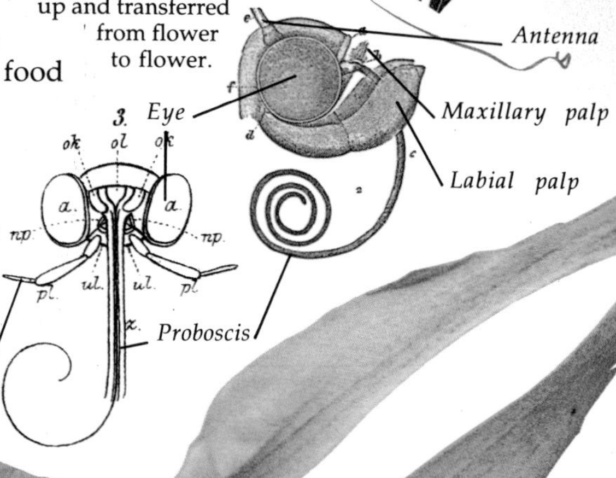

FINDING NECTAR *right*
The long proboscis of this hawk-moth seeks out nectar from flowers. During this probing, pollen is picked up and transferred from flower to flower.

Antenna

Eye

Maxillary palp

Labial palp

Head has the cerebral ganglion ("brain") inside. The eyes, antennae and sense organs called palps give the insect information about its environment

Proboscis

Labial palp

FACE-TO-FACE WITH A MOTH
An almost head-on view of the Indian Moon moth shows its antennae and front and middle legs. The antennae have very small sense organs that probably detect not only scent but changes in air pressure.

Since this moth does not feed as an adult it has no proboscis

This female Moon moth uses its antennae to select the correct food plant on which to lay its eggs

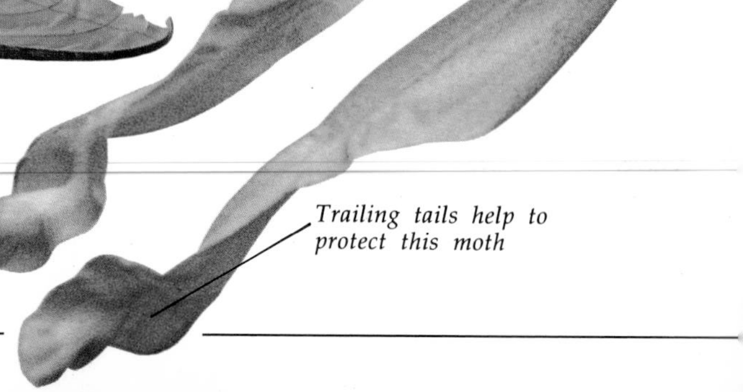

Trailing tails help to protect this moth

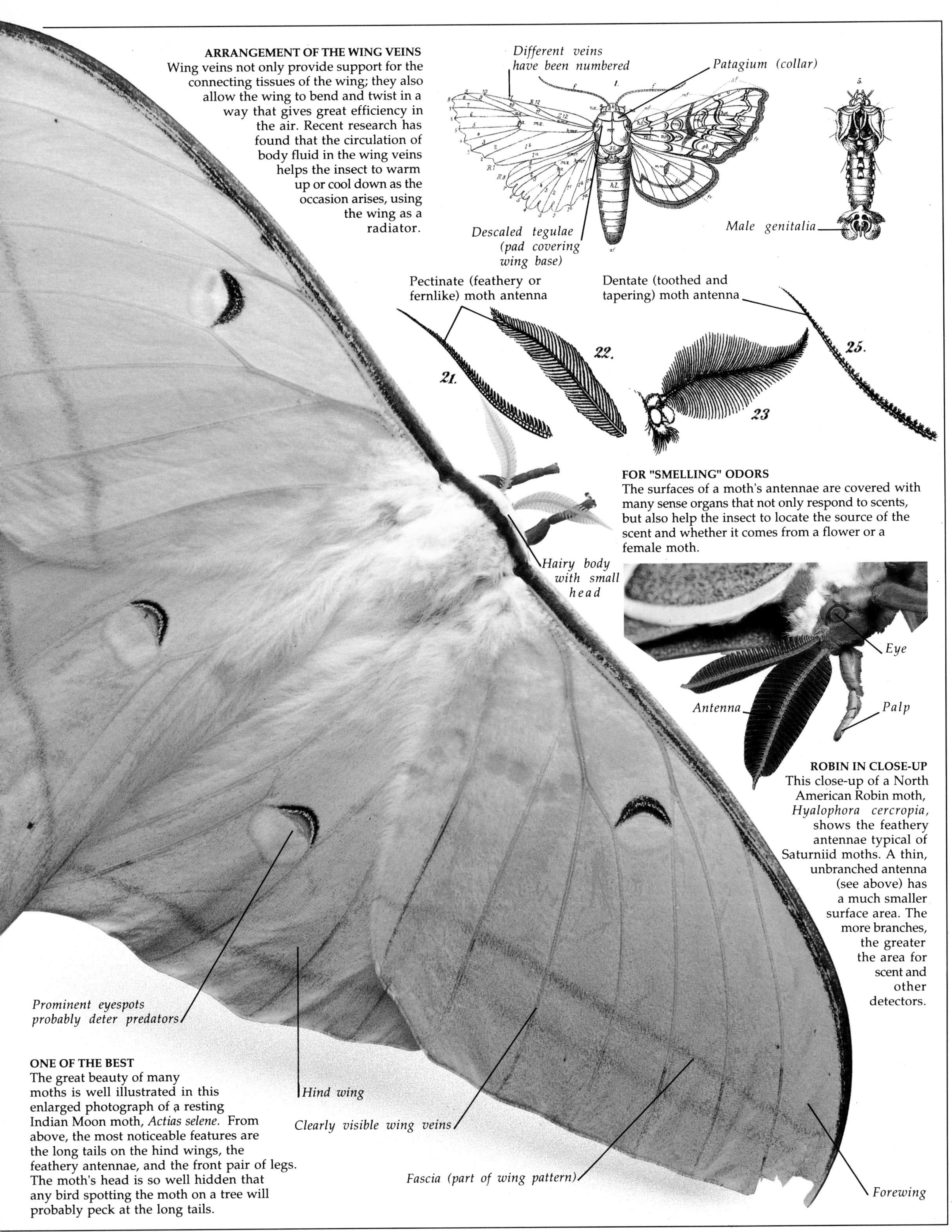

ARRANGEMENT OF THE WING VEINS
Wing veins not only provide support for the connecting tissues of the wing; they also allow the wing to bend and twist in a way that gives great efficiency in the air. Recent research has found that the circulation of body fluid in the wing veins helps the insect to warm up or cool down as the occasion arises, using the wing as a radiator.

FOR "SMELLING" ODORS
The surfaces of a moth's antennae are covered with many sense organs that not only respond to scents, but also help the insect to locate the source of the scent and whether it comes from a flower or a female moth.

ROBIN IN CLOSE-UP
This close-up of a North American Robin moth, *Hyalophora cercropia,* shows the feathery antennae typical of Saturniid moths. A thin, unbranched antenna (see above) has a much smaller surface area. The more branches, the greater the area for scent and other detectors.

ONE OF THE BEST
The great beauty of many moths is well illustrated in this enlarged photograph of a resting Indian Moon moth, *Actias selene.* From above, the most noticeable features are the long tails on the hind wings, the feathery antennae, and the front pair of legs. The moth's head is so well hidden that any bird spotting the moth on a tree will probably peck at the long tails.

Cocoons

MOST MOTHS SPIN A COCOON. This silken case encloses the caterpillar as it pupates and the inactive pupa while it is developing. Some species include stinging hairs from the last caterpillar skin, or bits of plant material, in the cocoon as added protection or camouflage. The development of cocoons reaches its peak in the silk moths (pp. 40-41), whose cocoons are made up of a single thread (sometimes about half a mile in length), wrapped around and around many times. When the adult moth is ready to emerge, it has to force its way out of the cocoon. This can be difficult, since the cocoon is often very hard. Some moths have a filelike organ with which to cut their way out; others produce a liquid that softens the walls. Many caterpillars also spin silken webs to protect themselves while they are feeding, but these are not true cocoons.

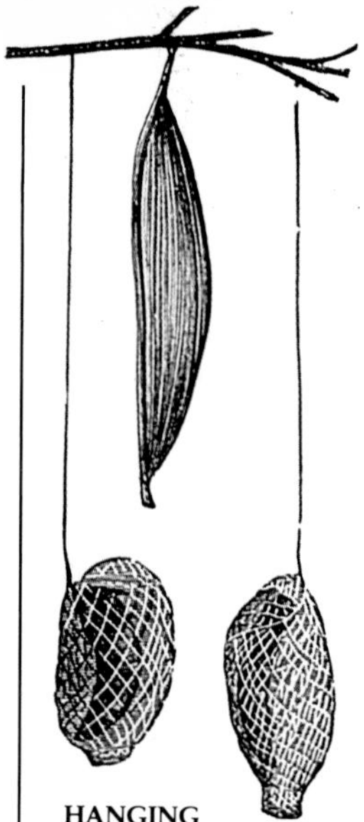

HANGING BY A THREAD
Some species suspend their cocoons from a long silken thread - an added protection against predatory insects.

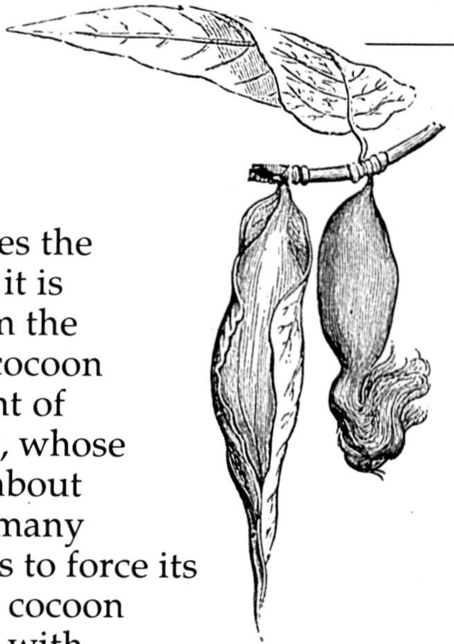

IN DISGUISE
These silken moth cocoons look very much like a part of the plant from which they hang, safe from all but the most sharp-eyed predators.

"WOODEN" COCOON
This Green Silverlines moth, *Bena fagana* (Europe), has just emerged. Its cocoon included bits of bark chewed off by the caterpillar, to give strength and camouflage.

Hard surface of cocoon, reinforced by fragments of bark

Distinctive wing pattern shows how the Green Silverlines gets its name

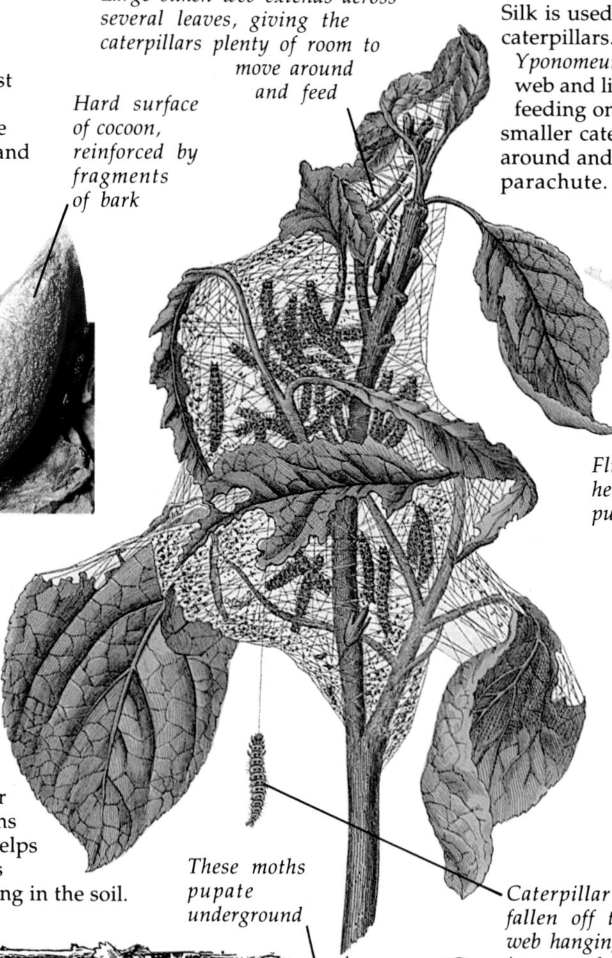

Large silken web extends across several leaves, giving the caterpillars plenty of room to move around and feed

CATERPILLAR SHELTER
Silk is used extensively by caterpillars. Many species of *Yponomeuta* moths spin a protective web and live under it in a group, feeding on the plant. Some of the smaller caterpillars may get blown around and use the silk like a parachute.

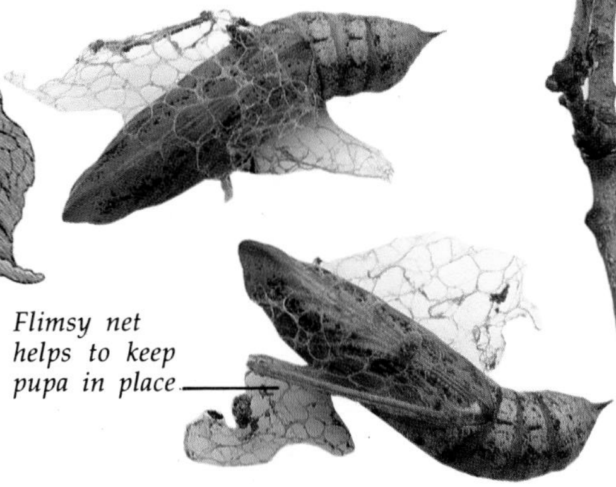

Flimsy net helps to keep pupa in place

UNDERGROUND NET
The pupae of the Silver-striped Hawkmoth, *Hippotion celerio* (Europe, Africa & Asia; also p. 19), have a very flimsy cocoon, consisting of a few strands of silk woven into a net.

BURIED ALIVE
Hawkmoths are one of the moth families that pupate below ground. The caterpillar makes a small cavity and spins silk around the walls. This helps to protect it from the damp as well as from animals burrowing in the soil.

These moths pupate underground

Caterpillar that has fallen off the web hanging on its own thread of silk

WILD SILK PRODUCERS
The Oak Silk moth, *Antheraea harti* (Asia; see pp. 62-63), is one of a number of species of wild silk moths in the family Saturniidae. Its caterpillars feed on oak leaves, among which they also spin their cocoons. These moths are related to the Chinese silk moths that produce most commercial silk.

THE COCOON TAKES SHAPE
The Oak Silk moth caterpillars are beginning to spin among the leaves near the top of the plant, and already the cocoon is starting to form. Eventually they will spin large quantities of silk to make their cocoons.

COMPLETED COCOON
The finished cocoon of the Oak Silk moth shows clearly how the leaves are bound together by a dense network of silk threads. These wild silk moth cocoons are used commercially but not on the scale of the silkworm, *Bombyx mori* (Asia; pp. 40-41).

Silk moths

SILK IS PRODUCED by most moth caterpillars; the finest quality silk is made by species of moths in the families Saturniidae and Bombycidae and, in particular, by the caterpillars of the large white moth, *Bombyx mori* (Asia), popularly known as the silkworm. According to Chinese legend, silk fiber was first discovered in about 2700 B.C., but for centuries the methods used to produce silk commercially were kept a well-guarded secret - the export of silkworms or their eggs out of China was punishable by death. Eventually silkworm eggs, and the seeds of the mulberry trees the caterpillars feed on, were smuggled out of China, supposedly hidden in a walking stick.

Silk continued to command high prices in Europe - even after the Arabs had introduced silkworms into Spain, and silk-weaving centers had been started in Italy. Today the silkworm has become so domesticated that it no longer occurs in the wild.

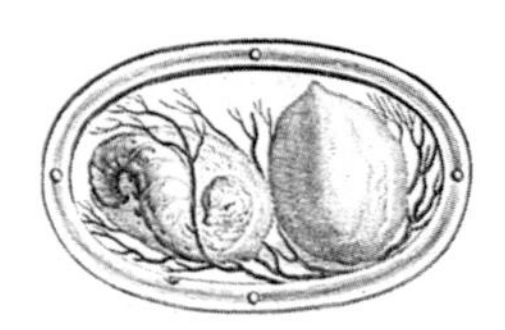

Illustrations from *Vermis sericus*, a popular 17th-century book on silk moths

SILKEN GOWNS
For many years, silk has been a highly prized material for luxurious wedding dresses and evening gowns.

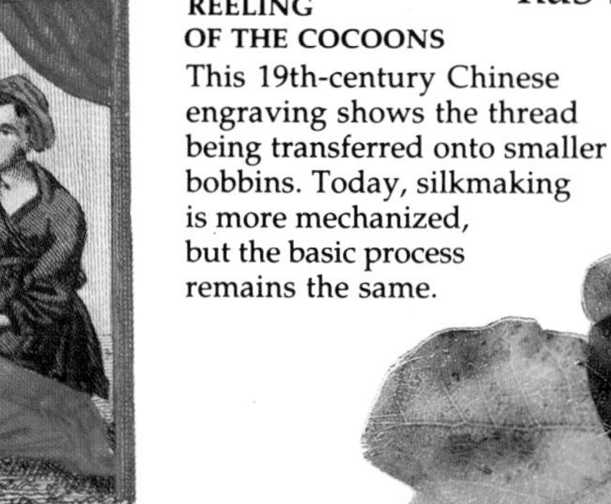

REELING OF THE COCOONS
This 19th-century Chinese engraving shows the thread being transferred onto smaller bobbins. Today, silkmaking is more mechanized, but the basic process remains the same.

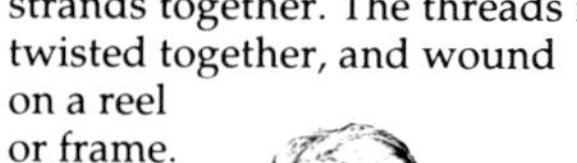

UNWINDING THE THREAD
From its origins in China (above) to 17th-century Europe (below), the methods used to produce silk changed little. The insects inside the cocoons were killed in boiling water before they could hatch and break the thread of silk. The hot water also had the effect of dissolving the gumlike substance holding the strands together. The threads from several cocoons were then caught up, twisted together, and wound on a reel or frame.

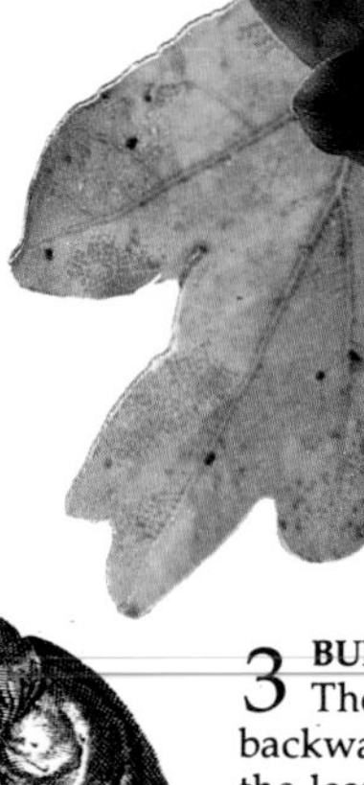

3 BUILDING UP THE WALLS
The caterpillar has worked backward and forward between the leaves, making the cocoon thicker. All the time the silk is being forced out through the caterpillar's spinneret.

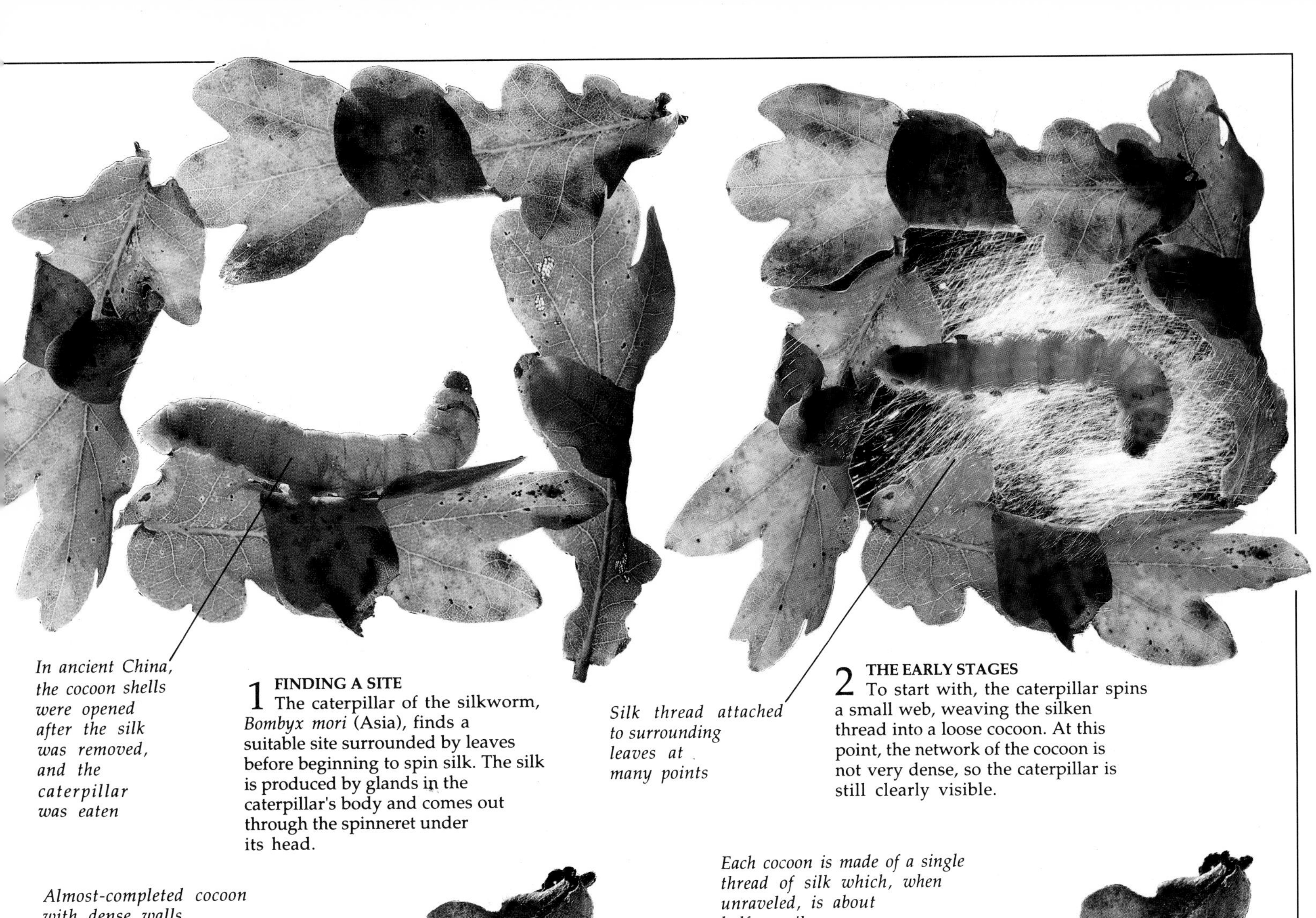

In ancient China, the cocoon shells were opened after the silk was removed, and the caterpillar was eaten

1 FINDING A SITE
The caterpillar of the silkworm, *Bombyx mori* (Asia), finds a suitable site surrounded by leaves before beginning to spin silk. The silk is produced by glands in the caterpillar's body and comes out through the spinneret under its head.

Silk thread attached to surrounding leaves at many points

2 THE EARLY STAGES
To start with, the caterpillar spins a small web, weaving the silken thread into a loose cocoon. At this point, the network of the cocoon is not very dense, so the caterpillar is still clearly visible.

Almost-completed cocoon with dense walls of silk

Each cocoon is made of a single thread of silk which, when unraveled, is about half a mile (805 m) long

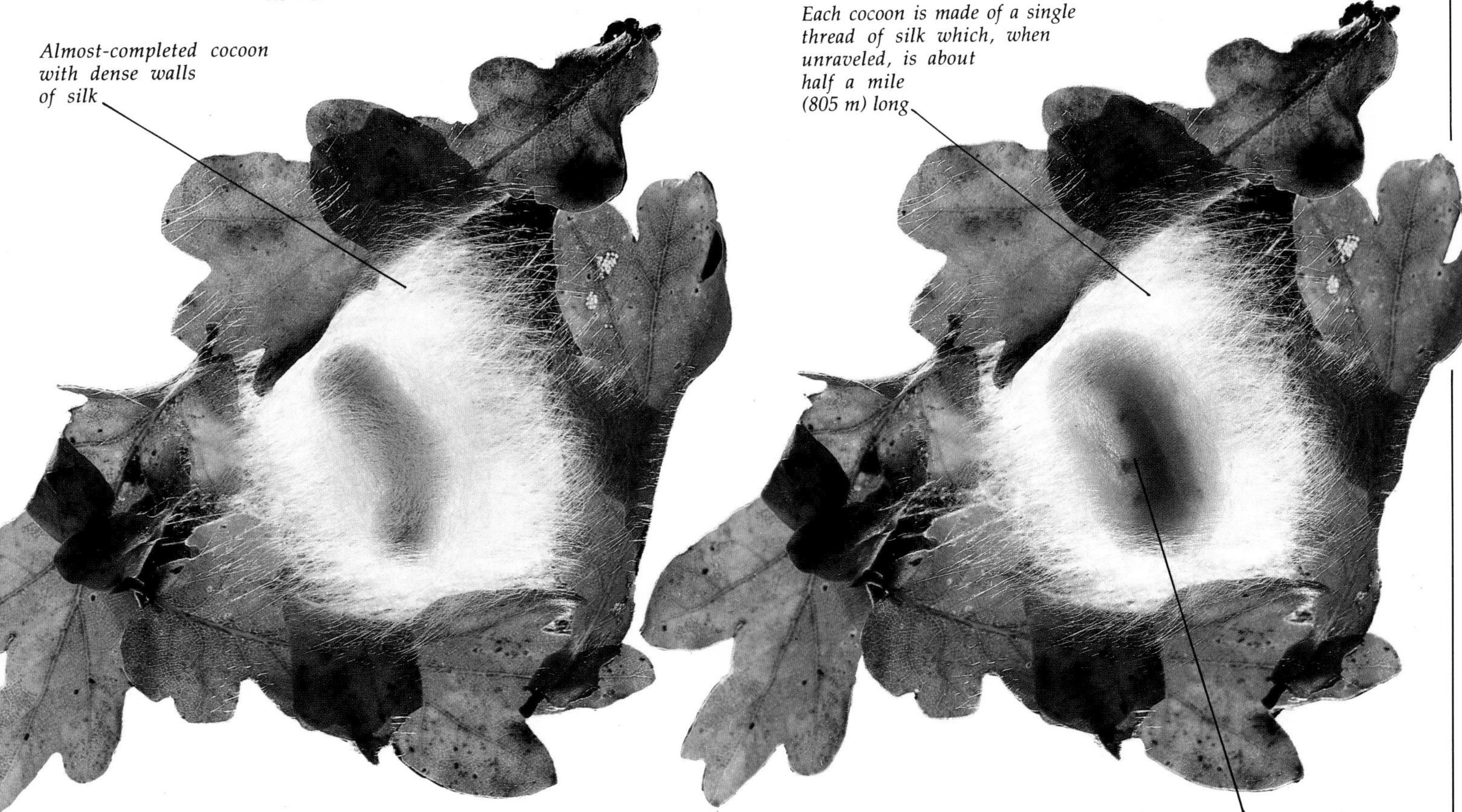

4 INCREASING THE DENSITY
The thickness of the silk layer increases and the cocoon will now keep most parasites and predators away from the caterpillar.

5 A SAFE HAVEN
The cocoon is now strong enough to protect the caterpillar fully as it starts to pupate and, eventually, as it changes into a moth.

Fully protected caterpillar can now begin to pupate

Temperate moths

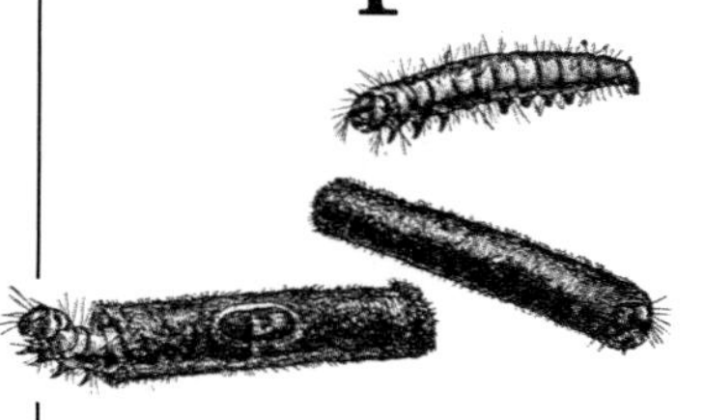

TEMPERATE MOTHS, like butterflies from the same regions (pp. 28-29), have to be able to survive the cold winter months. Some temperate moths remain at the egg stage, and others pass the winter as caterpillars, perhaps concealed inside the stem of a plant. Many more survive as a pupa, which in some species is further protected in a cocoon (pp. 38-39). In the temperate areas of Europe, Asia and North America, a moth's life cycle is synchronized with the spring and summer months when there are plenty of grasses and flowers. The greatest variety of temperate moths are found during warm summer nights, when they can be seen at windows, or flying around some other source of light. On a moonlit night it may just be possible to see them sipping the nectar from flowers. Although most temperate moths fly only by night, certain species are active in the daytime (pp. 48-49).

Female of species is white throughout; in the male, upper surface of hind wing is yellow

MISNAMED MOTH *above*
Despite its name, the Salt-marsh moth, *Estigmene acrea*, occurs in a variety of habitats throughout North America and Mexico.

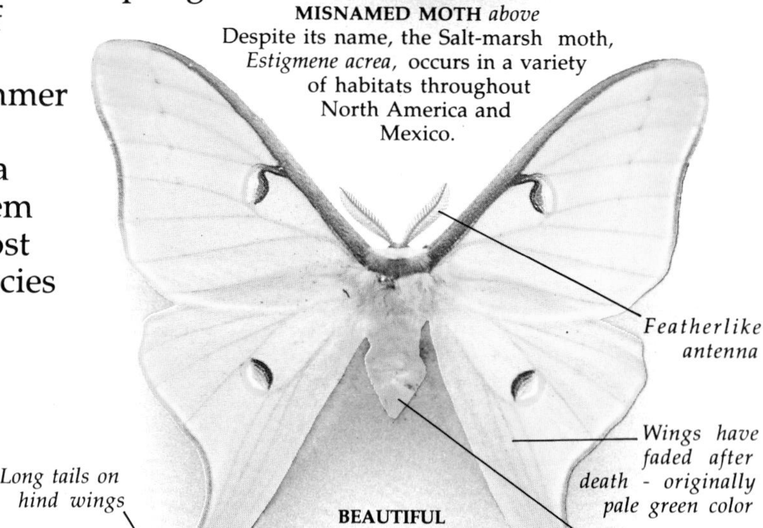

BEAUTIFUL SILK MOTH
The spectacular Luna moth, *Actias luna*, is found only in North America. In recent years its numbers have decreased due to pollution and insecticides. Like the Indian Moon moth (pp. 36-37), it is often bred in captivity.

Bright pattern warns predators that the moth is poisonous

POISONOUS TIGER *above*
Like other tiger moths, the Virgin Tiger moth, *Grammia virgo* (N. America), is avoided by birds because of its distasteful body fluids.

HOLLOWING A HOME
The caterpillar of the Locust-bean moth, *Ectomyelois ceratoniae* (worldwide), pupates in hollowed-out thistle stems.

PALER TIGER MOTH *right*
This day-flying Zuni Tiger, *Arachnis zuni*, can be found in the southwestern United States and Mexico. Like the Virgin Tiger (above), its color gives warning

Dull-colored forewings

When disturbed, the red hind wings are flashed to confuse predators

STARING EYES
The Eyed Hawkmoth, *Smerinthus ocellata* (Europe & Asia), gets its name from the eyespots on its hind wings. If the moth is disturbed, it moves its forewings to reveal two large, staring eyes.

Eyespots scare enemies away

FRUIT EATER *left*
Because the caterpillar of the Codling moth, *Cydia pomonella* (worldwide), feeds on apples and pears, it is often called the apple maggot.

FLASHY HIND WINGS
Underwing moths are found in Europe, Asia, and America. The Bronze Underwing, *Catocala cara*, is a North American species whose range covers Canada to Florida.

PINES ARE FINE
By feeding on pine trees, the caterpillar of the Resin-gall moth, *Petrova resinella* (Europe, N. America & Asia), causes lumps of resin to be released.

ELEPHANT, HAWK, OR MOTH?
The attractive Elephant Hawkmoth, *Deilephila elpenor* (Europe & Asia), gets its name from the trunklike shape of the caterpillar.

LOOPING ALONG *above*
The caterpillar of the Swallowtail moth, *Ourapteryx sambucaria* (Europe & Asia), moves its body by a series of "looping" actions. In the United States similar caterpillars are called inchworms or measuring worms.

LIKE A BROKEN TWIG
The Bufftip, *Phalera bucephala* (Europe & Asia), manages to look exactly like a twig when it is at rest. The yellow areas on the end of its wings and around its head explain why it is so named.

SKULL-LIKE MARKINGS
One of the most interesting temperate moths is the Death's-head Hawkmoth, *Acherontia atropos* (Europe, Asia & Africa). Not only does it have a skull-like pattern on its body, but it also squeaks if disturbed. The caterpillar feeds on the leaves of the potato plant. The adult sometimes steals honey from beehives (also pp. 14-15).

BIG IN EUROPE
The largest European moth, the Great Peacock moth, or Viennese Emperor, *Saturnia pyri,* belongs to the same family as the giant silk moths. It can be found mostly in southern Europe and western Asia.

FADED BEAUTY
The Large Emerald, *Geometra papilionaria* (Europe & Asia), has "looper" or inchworm caterpillars (see opposite page) that hibernate on twigs.

COMPARE THE HIND WINGS
Closely related to the Bronze Underwing (opposite page), the Clifden Nonpareil, *Catocala fraxini,* is found throughout Europe and Asia. Its beautiful hind wings have a quite different pattern than its duller fore-wings.

A DRINKING HABIT
Philudoria potatoria is known as the Drinker because of the caterpillar's habit of drinking dew off the grasses on which it feeds. It is found in damp places throughout Europe, and in Asia as far as Japan.

Exotic moths

BECAUSE MOST MOTHS HIDE DURING THE DAY, it is easy to forget that many of them are just as colorful and exotic-looking as butterflies. This is especially true of moths that live in the tropical and subtropical regions of Africa, Asia, Australia, and South and Central America. Here the nights are full of the most amazing-looking moths, many of which have never been seen except by the collectors who seek them out. During the daytime most of these moths conceal themselves in the lush tropical vegetation. At night, photography of moths in the wild is difficult, and still in its early stages. It is therefore revealing to look at the great variety of color, pattern, and shape of the tropical species shown over the next four pages.

MOTH ROCKET
The curious-looking "rocket-like" tufts of the Noctuid moth, *Epicausis mithii* (Madagascar), are probably scales that emit scent.

POTATO PEST
A destroyer of sweet potatoes, the Sweet Potato Hornworm, *Agrius cinqulatus,* is found in South, Central, and North America.

TIGER MOTH BIPLANE
In the 1920s, the British aircraft maker Geoffrey de Havilland named an entire range of two-seater light planes after powerful flying insects such as the Tiger moth.

STRONG FLYER
Like most hawkmoths, the beautiful Verdant Hawkmoth, *Euchloron megaera* (Africa), has a thick body and is a powerful flyer.

BUTTERFLY SHAPE
The shape of its wings make this Chalcosine moth, *Agalope caudata* (Japan), look more like a butterfly than a moth.

ABORIGINE FOOD *above*
This Hepialid moth, *Charagia mirabilis,* comes from Australia. Hepialid caterpillars are collected and eaten by Australian aborigines.

DAY-FLYING BEAUTY *left*
Among the most attractive of the tropical moth families is the Uraniidae. This day-flying Uraniid, *Alcides aurora,* with its feathery hind wing, is found in New Guinea and the Solomon Islands.

MOTH MIMIC
The Pterothysanid moth, *Hibrildes norax* (Africa), flies by day and mimics a distasteful butterfly. To add to its camouflage, it even folds its wings over its back like a butterfly.

TRAILING TAILS *below*
One of the strikingly beautiful Saturniidae family, the Madagascan Moon moth, *Argema mittrei,* has large startling eyespots on its wings.

EAST AND WEST *right*
In this old engraving, two South American hawkmoths of the species *Xylopanes chiron* are seen with an Asian moth, *Euchromia polymena* (top).

MULTICOLORED BRILLIANCE *right*
Not surprisingly, the beautiful Jamaican Uraniid moth, *Urania sloanus,* is often mistaken for a butterfly. A closer look reveals the long slender moth antennae.

FLASHY TYPE
Like other Castniid moths, *Cyanostola hoppi* (C. America) is a day-flyer with bright "flash" colors on its hind wings. When the resting moth moves its wings, the sudden flash of color scares away would-be predators.

SHAPE BREAKER
When resting, this Noctuid moth from South America holds its hind wings slightly in front of its forewings. This will confuse predators by breaking up the moth's shape. The tufts or long scales are hair pencils (pp. 56-57) for the male moth to disperse its scent.

BIG MOTH, SMALL FAMILY
The Brahmaeidae are one of the smallest moth families, with only about 20 known species. This Brahmaeid moth, *Brahmaea wallichii,* with its wonderful swirling camouflage pattern, comes from Southeast Asia.

Continued on next page

KEEPING AN EYE OPEN
When this Saturniid, *Ludia dentata* (Africa), moves its forewings, they reveal "eyes" that usually are meant to scare off predators.

TRAILING WINGS
The intricate pattern of scales and the curious wing shape of the Tailed Saturniid moth, *Copiopteryx decerto* (S. America), make this a very distinctive moth.

ADDED PROTECTION
The Pericopine moth, *Chetone phyleis* (S. America), mimics a distasteful *Heliconius* butterfly (pp. 56-57).

NEWLY EMERGED MOTHS
When it emerges from the pupa, a moth's crumpled wings give it an unreal appearance (pp. 24-25).

WELL CAMOUFLAGED
Few large moths are better at blending into the background than the Saturniid, *Loxolomia serpentina* (S. America).

LEAFLIKE CAMOUFLAGE
Although little is known about the Midilid moth, *Eupastrana fenestrata* (S. America), it seems to have an effective decaying-leaf camouflage.

WARNING WINGS *below*
If the West African Eupterotid moth, *Acrojan rosacea,* is disturbed, it flashes its hind wings to startle any would-be predator.

NASTY TO EAT
The Chalcosiine moth, *Campylotes kotzschi* (India), is avoided by birds, who sense from its warning colors that it is unpleasant to eat.

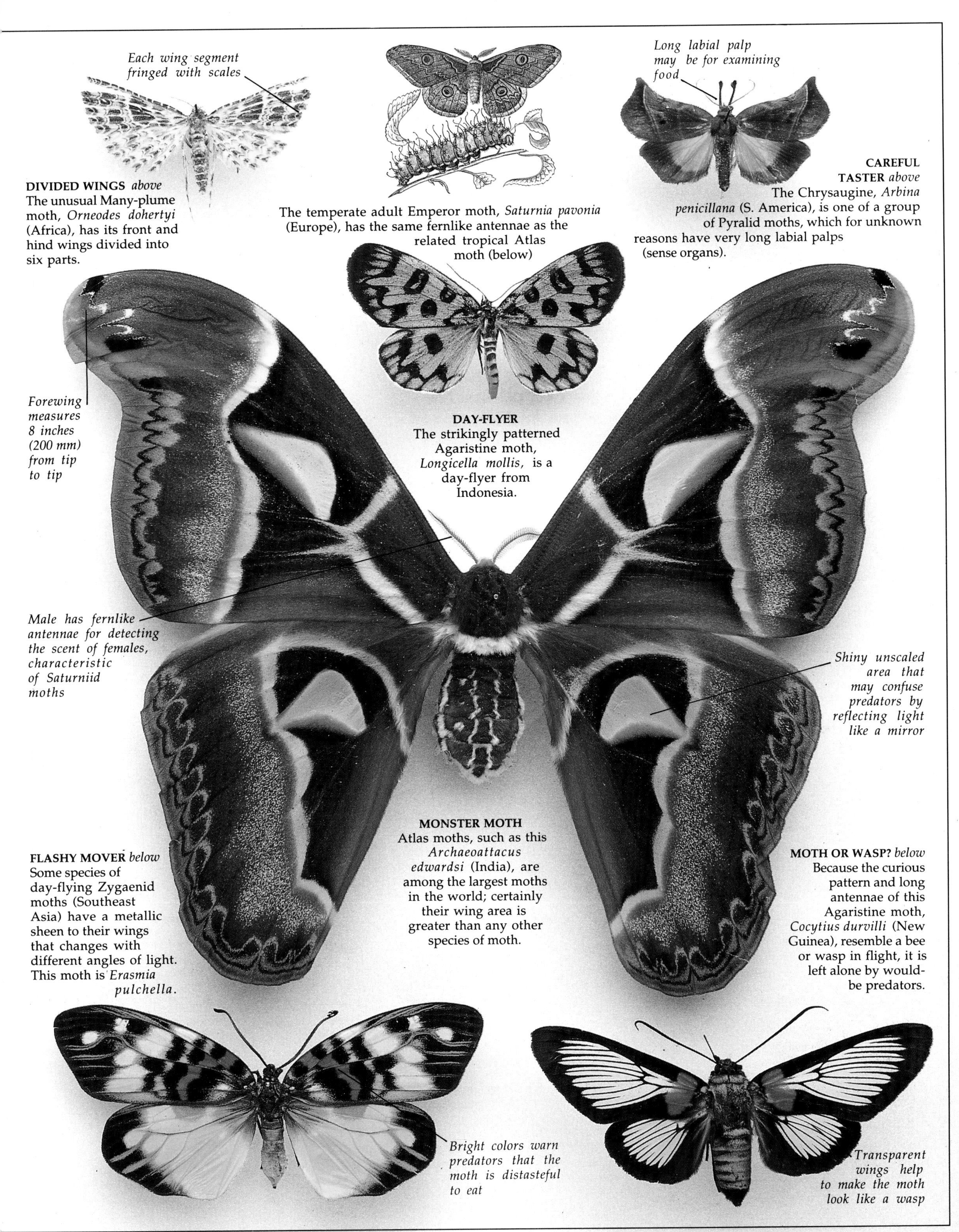

DIVIDED WINGS *above*
The unusual Many-plume moth, *Orneodes dohertyi* (Africa), has its front and hind wings divided into six parts.

The temperate adult Emperor moth, *Saturnia pavonia* (Europe), has the same fernlike antennae as the related tropical Atlas moth (below)

CAREFUL TASTER *above*
The Chrysaugine, *Arbina penicillana* (S. America), is one of a group of Pyralid moths, which for unknown reasons have very long labial palps (sense organs).

DAY-FLYER
The strikingly patterned Agaristine moth, *Longicella mollis*, is a day-flyer from Indonesia.

MONSTER MOTH
Atlas moths, such as this *Archaeoattacus edwardsi* (India), are among the largest moths in the world; certainly their wing area is greater than any other species of moth.

FLASHY MOVER *below*
Some species of day-flying Zygaenid moths (Southeast Asia) have a metallic sheen to their wings that changes with different angles of light. This moth is *Erasmia pulchella*.

MOTH OR WASP? *below*
Because the curious pattern and long antennae of this Agaristine moth, *Cocytius durvilli* (New Guinea), resemble a bee or wasp in flight, it is left alone by would-be predators.

Day-flying moths

MOTHS ARE USUALLY THOUGHT of as creatures of the night. While this is true of the majority of the 150,000 species, there are a large number that are day-flyers. Many moths will fly by day if disturbed, but the ones illustrated on these two pages are specialist day-flyers. Flying during the day means that in many ways they behave like butterflies, but the structure of their bodies, particularly the way in which their front and hind wings link together, shows that they are moths. Many of them can be seen around flowers and are often mistaken for butterflies. But their wing shapes are usually different and their antennae do not generally end in a club (pp. 6-7). There are, however, always exceptions: the Zygaenid moths have swollen antennae, and species of Urania moths have butterfly-shaped wings, although their antennae are slender and mothlike. Some of these insects belong to families in which the majority are night-flyers; others, like the Zygaenidae, are mainly day-flying moths. Day-flyers include many interesting species such as the Hummingbird Hawk-moth, which hovers in front of flowers and sucks out nectar with its long proboscis. Many day-flying moths are also brightly colored and noticeable.

These two Pyralids are (top) the Gold Spot, *Pyraustra purpuralis* (Europe & Asia), and the Small Magpie, *Eurrhypara hortulata*, (Europe & Asia).

STRIPED BODY
The wings of this Euchromiid, *Euchromia lethe* (Africa), are not as decorative as those of some moths, but it has brightly colored bands across its body. Many moths have striped bodies. Some mimic wasps; others are even more vividly colored. This species is sometimes found on imported bananas.

Butterfly-shaped wings and slender antennae, together with its bright colors, make this a particularly striking moth

SWEET TEMPTATIONS
To encourage butterflies and moths to come to your garden you should grow plants that attract them with nectar and perfume. Some plants, like species of Hebe (right), buddleia, and aster, are particularly attractive to butterflies and moths.

"Furry" bee-like striped abdomen

ROOT BORER
The Big-root Borer, or Bee moth, *Melittia gloriosa* (N. America), bears a striking similarity to a bee. This is even more obvious from its flight and its behavior in the field. This species feeds on the roots of squashes and gourds.

POISON EATER
Sloan's Uraniid moth, *Urania sloanus* (Jamaica), has unusual feeding habits. The caterpillars feed on species of plants that are poisonous to most animals but not to this insect, which derives protection from the poison.

SLOW FLYER
This Pericopine, *Gnophaea arizonae* (N. & C. America), is a slow-flying moth often found in great numbers in meadows over 8,200 ft (2,500 m) above sealevel. Its slow flight, bold pattern, and daytime activity suggest that predators avoid it because it is distasteful.

METALLIC MOTH
This Geometrid, *Milionia paradisea* (Papua New Guinea), has metallic colors that catch the sun as it flies. The popular idea of a brown moth is not borne out by this species.

PREDATORS, BEWARE!
This colorful Syntomid, *Syntomis phegea* (Europe & Asia), is common around flowers in the warmer parts of Europe and Asia. A distasteful species, it is avoided by birds.

REFLECTOR
The metallic appearance of this Geometrid, *Milionia exultans* (Bismarck Archipelago), is caused by scales on the wings, which have ridges along them that catch the light.

VARYING PATTERNS
The Variable Burnet, *Zygaena ephialtes* (Europe & Asia), has different forms of wing pattern, with either red or yellow spots, making this day-flying moth very popular with collectors.

COLLECTORS' FAVORITE
The Provence Burnet, *Zygaena occitanica* (Europe), also has variable wing patterns and is equally popular with collectors.

HUMMINGBIRD?
Most reports of "hummingbirds" around flowers in Europe or Asia stem from sightings of the Hummingbird Hawkmoth, *Macroglossum stellatarum* (Europe, Asia & N. Africa).

HAWKMOTH IN ACTION
This illustration of a Hummingbird Hawkmoth is from a drawing by collector Moses Harris and was produced for a new edition of his classic 18th-century work *The Aurelian* (p. 58). On the ground is the nocturnal Dot moth, *Melanchra persicariae* (Europe & Asia).

EXOTIC AGARISTID
Agaristids, like *Exsula dentatrix* (Asia), are mostly tropical moths with a few representatives in North America but none in Europe. Many are day-flyers and brightly colored, often with a basic orange and black pattern.

WARNING PATTERNS
A number of moths have developed the bright colors of this Pericopine, *Ephestia melaxantha* (S. America). This is probably to warn predators. Although most of these moths are day-flying, little is known of their habits.

Migration and hibernation

BIRD MIGRATION HAS BEEN KNOWN about for hundreds if not thousands of years, but the migration of butterflies and moths is a relatively recent discovery. Unlike birds, most butterflies migrate in one direction only - from the place where they were born to a new area. There are several possible reasons for this - to avoid overpopulation; to find a new home when a temporary habitat such as agricultural land is destroyed; or to respond to the changing seasons. While birds tend to migrate at the onset of bad weather, butterflies and moths often migrate when the weather improves. For example, some species move north from North Africa and southern Europe as new plant growth becomes available for egg laying. It may not be a direct flight - they may breed on the way.

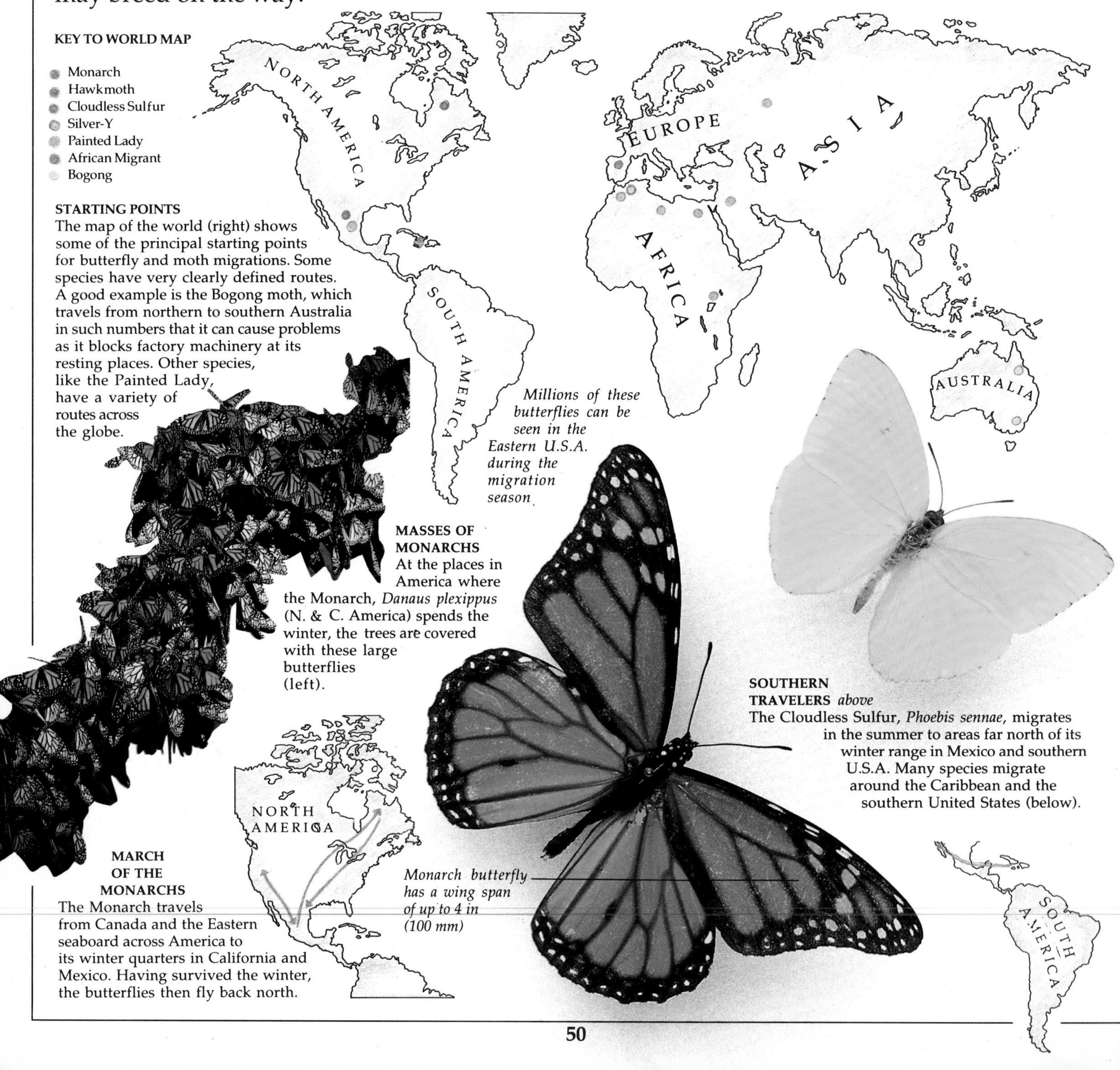

KEY TO WORLD MAP

- Monarch
- Hawkmoth
- Cloudless Sulfur
- Silver-Y
- Painted Lady
- African Migrant
- Bogong

STARTING POINTS
The map of the world (right) shows some of the principal starting points for butterfly and moth migrations. Some species have very clearly defined routes. A good example is the Bogong moth, which travels from northern to southern Australia in such numbers that it can cause problems as it blocks factory machinery at its resting places. Other species, like the Painted Lady, have a variety of routes across the globe.

Millions of these butterflies can be seen in the Eastern U.S.A. during the migration season

MASSES OF MONARCHS
At the places in America where the Monarch, *Danaus plexippus* (N. & C. America) spends the winter, the trees are covered with these large butterflies (left).

SOUTHERN TRAVELERS *above*
The Cloudless Sulfur, *Phoebis sennae*, migrates in the summer to areas far north of its winter range in Mexico and southern U.S.A. Many species migrate around the Caribbean and the southern United States (below).

Monarch butterfly has a wing span of up to 4 in (100 mm)

MARCH OF THE MONARCHS
The Monarch travels from Canada and the Eastern seaboard across America to its winter quarters in California and Mexico. Having survived the winter, the butterflies then fly back north.

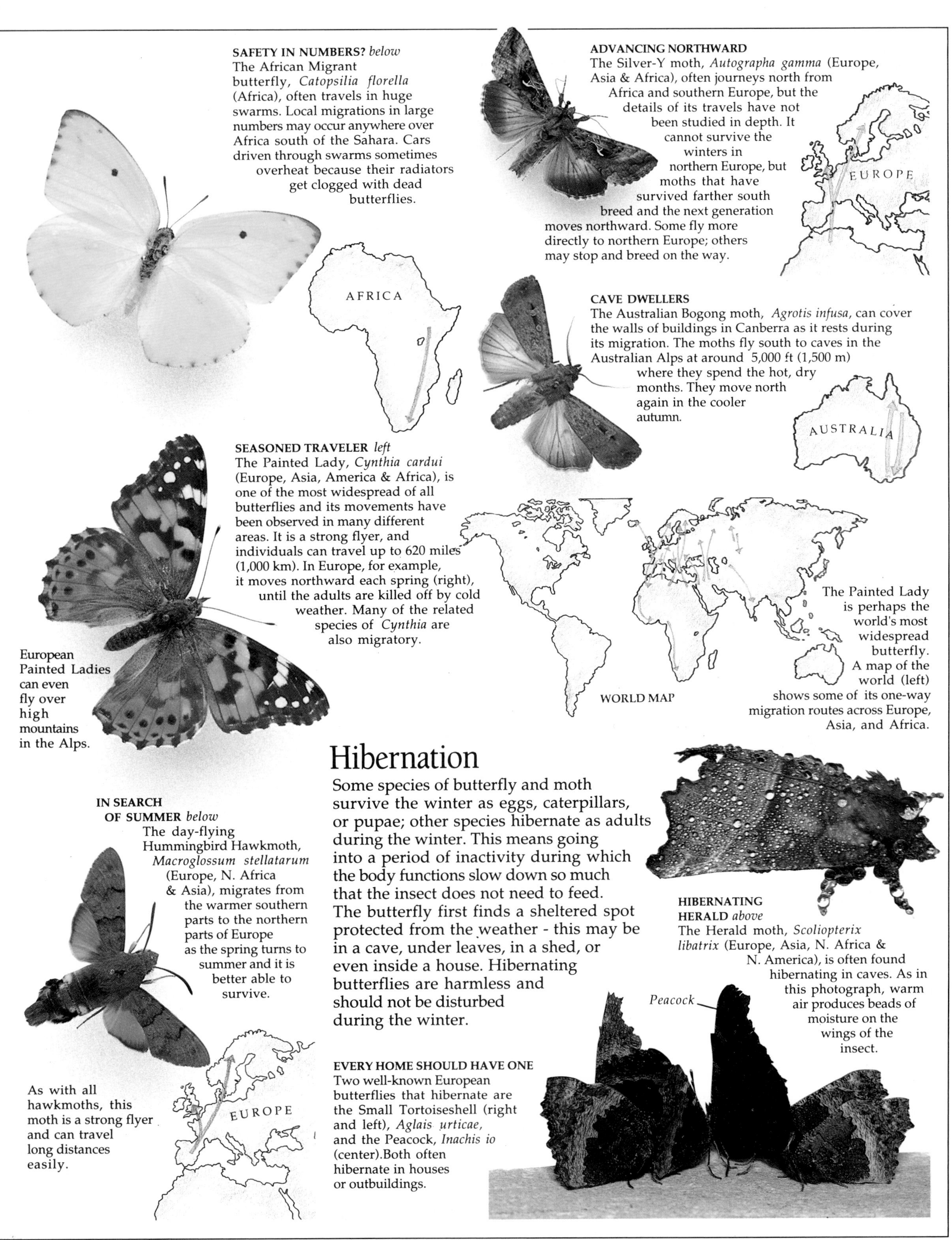

SAFETY IN NUMBERS? *below*
The African Migrant butterfly, *Catopsilia florella* (Africa), often travels in huge swarms. Local migrations in large numbers may occur anywhere over Africa south of the Sahara. Cars driven through swarms sometimes overheat because their radiators get clogged with dead butterflies.

ADVANCING NORTHWARD
The Silver-Y moth, *Autographa gamma* (Europe, Asia & Africa), often journeys north from Africa and southern Europe, but the details of its travels have not been studied in depth. It cannot survive the winters in northern Europe, but moths that have survived farther south breed and the next generation moves northward. Some fly more directly to northern Europe; others may stop and breed on the way.

CAVE DWELLERS
The Australian Bogong moth, *Agrotis infusa*, can cover the walls of buildings in Canberra as it rests during its migration. The moths fly south to caves in the Australian Alps at around 5,000 ft (1,500 m) where they spend the hot, dry months. They move north again in the cooler autumn.

SEASONED TRAVELER *left*
The Painted Lady, *Cynthia cardui* (Europe, Asia, America & Africa), is one of the most widespread of all butterflies and its movements have been observed in many different areas. It is a strong flyer, and individuals can travel up to 620 miles (1,000 km). In Europe, for example, it moves northward each spring (right), until the adults are killed off by cold weather. Many of the related species of *Cynthia* are also migratory.

European Painted Ladies can even fly over high mountains in the Alps.

The Painted Lady is perhaps the world's most widespread butterfly. A map of the world (left) shows some of its one-way migration routes across Europe, Asia, and Africa.

Hibernation

Some species of butterfly and moth survive the winter as eggs, caterpillars, or pupae; other species hibernate as adults during the winter. This means going into a period of inactivity during which the body functions slow down so much that the insect does not need to feed. The butterfly first finds a sheltered spot protected from the weather - this may be in a cave, under leaves, in a shed, or even inside a house. Hibernating butterflies are harmless and should not be disturbed during the winter.

IN SEARCH OF SUMMER *below*
The day-flying Hummingbird Hawkmoth, *Macroglossum stellatarum* (Europe, N. Africa & Asia), migrates from the warmer southern parts to the northern parts of Europe as the spring turns to summer and it is better able to survive.

As with all hawkmoths, this moth is a strong flyer and can travel long distances easily.

HIBERNATING HERALD *above*
The Herald moth, *Scoliopterix libatrix* (Europe, Asia, N. Africa & N. America), is often found hibernating in caves. As in this photograph, warm air produces beads of moisture on the wings of the insect.

EVERY HOME SHOULD HAVE ONE
Two well-known European butterflies that hibernate are the Small Tortoiseshell (right and left), *Aglais urticae*, and the Peacock, *Inachis io* (center).Both often hibernate in houses or outbuildings.

Shape, color, and pattern

MOTHS ARE AMONG the most colorful creatures in the world. Butterflies have been described as "flying flowers," but moths have unrivaled wing patterns and show more variation in the shape of their wings as well. Color and pattern play important roles in the lives of these insects. They may provide protection by means of camouflage (pp. 54-55), or they may help to advertise a moth's presence. By making the insect stand out, the colors may remind predators that the creature is distasteful and should be avoided; bright colors may also imitate a dangerous insect such as a wasp - another way of discouraging predators. On the other hand, striking colors may also help to attract a mate.

MOTH OR WASP?
The day-flying Hornet Clearwing moth, *Sesia apiformis* (America, Europe & Asia), looks like a hornet and even has a similar flight pattern. Few predators would be rash enough to risk attacking this insect.

SHARP DRESSER
The Noctuid *Apsarara radians* (Asia) has a thornlike wing pattern.

VANISHING OUTLINE
Markings disguise the shape of the African Noctuid *Mazuca strigicincta.*

DISAPPEARING TRICK
The Noctuid *Diphthera festiva* (C. & S. America) is almost invisible against a suitable background.

DELICATE PATTERNS
This Noctuid, *Baorisa hieroglyphica* (E. Asia), has lines and stripes that break up its wings.

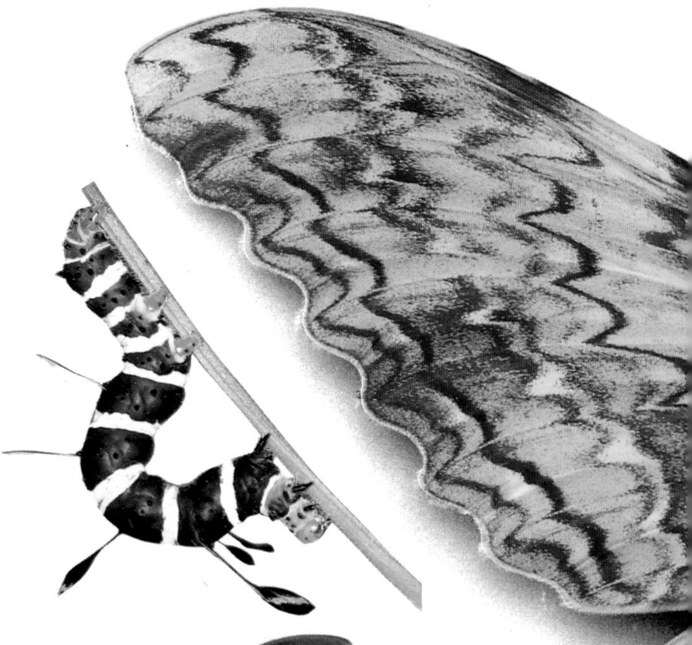

PUTTING OFF BIRDS
The caterpillar of the Alder moth, *Acronicta alni* (Europe), looks like a bird dropping when it is small. In its later stages it takes on a more aggressive appearance as the white markings are replaced by orange. Hairs along its back give it its unusual shape.

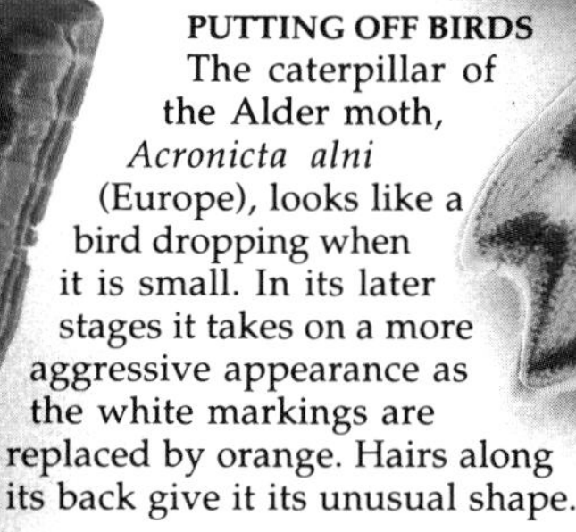

False "eyes"

Flagella

FEARSOME FOE
The caterpillar of the Puss moth, *Cerura vinula* (Europe & Asia), uses red markings and false "eyes" in its aggression display. It also waves the threadlike extensions (flagella) on its tail.

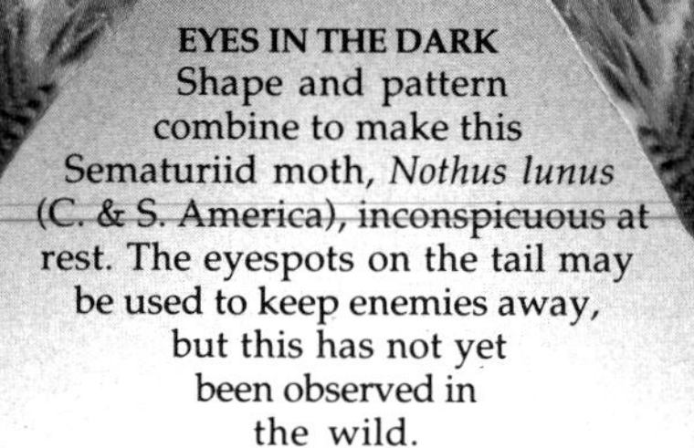

EYES IN THE DARK
Shape and pattern combine to make this Sematuriid moth, *Nothus lunus* (C. & S. America), inconspicuous at rest. The eyespots on the tail may be used to keep enemies away, but this has not yet been observed in the wild.

PEARL OF THE ORIENT
This Pyralid moth, *Margaronia quadrimaculalis* (E. Asia), has pearl-like white wings, broken up by a brown pattern.

SPOTTED WINGS
The forewings of the Thyridid *Rhodoneura limatula* (Madagascar) provide effective camouflage.

CONFUSION OF COLORS
Cerace xanthocosma (Japan), with its spots and swirls of color, is aptly named the Kaleidoscope moth.

MOTH IN WASP'S CLOTHING
This Pyralid, *Glyphodes militans* (E. Asia), gains some protection because its body looks like a wasp's.

RED FOR DANGER
The Arctiid *Composia credula* (W. Indies) has a pattern of red and black, warning predators that it is distasteful.

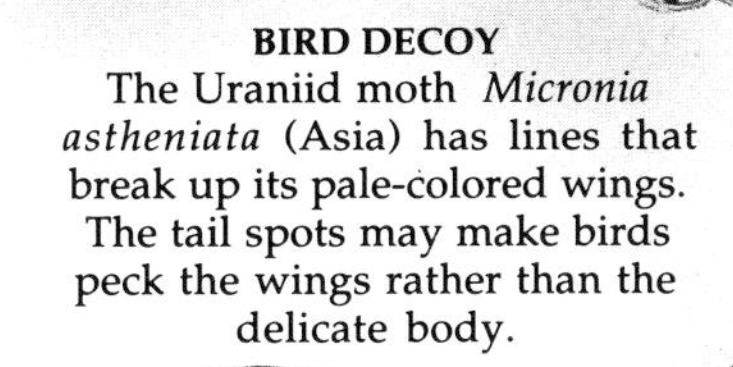

BIRD DECOY
The Uraniid moth *Micronia astheniata* (Asia) has lines that break up its pale-colored wings. The tail spots may make birds peck the wings rather than the delicate body.

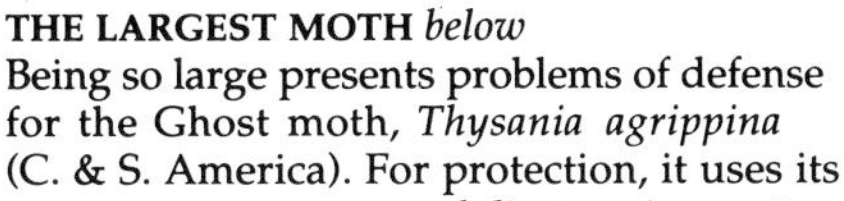

THE LARGEST MOTH *below*
Being so large presents problems of defense for the Ghost moth, *Thysania agrippina* (C. & S. America). For protection, it uses its delicate wing pattern as camouflage against tree trunks.

Wing span up to 12 in (300 mm)

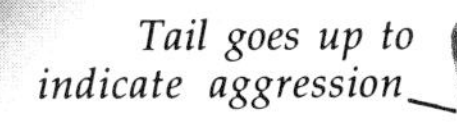

Tail goes up to indicate aggression

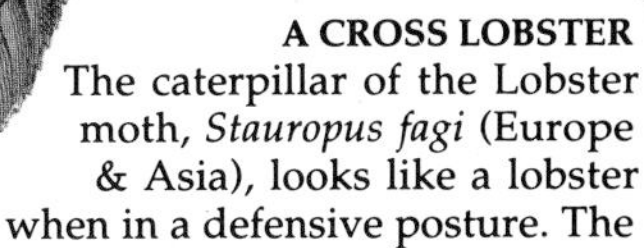

A CROSS LOBSTER
The caterpillar of the Lobster moth, *Stauropus fagi* (Europe & Asia), looks like a lobster when in a defensive posture. The head goes up and back, and the tail is held up and forward. Like the Puss moth caterpillar, it also waves its tail.

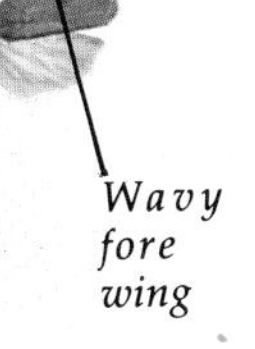

WAVY LINES
In the Tortricid *Acleros emargana* (N. America, Europe & Asia) the principle of camouflage is to avoid straight lines in its pattern - the forewing is particularly wavy. The forewings conceal the white hind wings at rest and the moth "vanishes" against its background.

Wavy fore wing

"LONG-TAILED" MOTH
Himantopterus marshalii, a Zygaenid moth from Africa, is given its distinctive shape by long hind wings that look like tails. The moth flutters above the grass with the hind wings floating up and down behind.

Distinctive hind wing, like long tail

Camouflage

All wild creatures have ways of protecting themselves from their enemies. For edible butterflies and moths a successful way of avoiding an early death is to "disappear" into their surroundings. They may do this by mimicking another object, or they may take on the patterns and colors of local trees, rocks, or leaves. Because they are especially vulnerable in daylight hours, many caterpillars and resting moths have perfected the art of concealment. Butterflies, which are active by day, and which usually rest with their wings together over their backs, have adopted other forms of camouflage. Some forest butterflies rest like moths with their wings spread out; other species disguise themselves as either living or decaying leaves. The butterfly that has perfected this clever form of camouflage is the Indian Leaf butterfly, truly a master of disguise.

DEADLY ENEMY *left*
One of the main reasons why many moths and butterflies camouflage themselves is to escape from predatory birds.

Leaf butterfly at rest on stem

Brown underside of Leaf butterfly

Orange and blue upperside of Leaf butterfly

INDIAN LEAF-TRICK
The most dramatic example of butterfly camouflage is the Indian Leaf butterfly, *Kallima inachus* (S.E. Asia). At rest, the butterfly looks remarkably like a decaying leaf on a stem. It frequently rests on the ground in leaflitter, where it becomes virtually invisible.

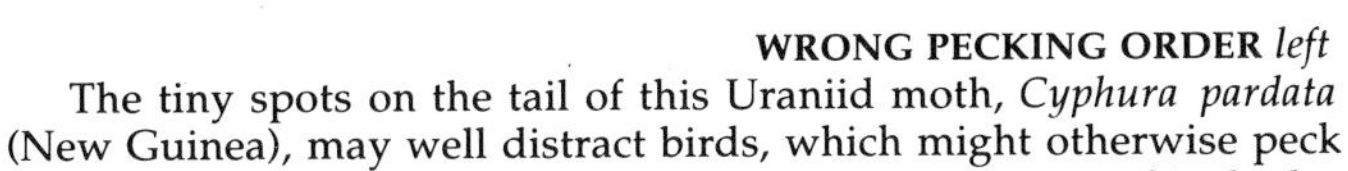

WRONG PECKING ORDER *left*

The tiny spots on the tail of this Uraniid moth, *Cyphura pardata* (New Guinea), may well distract birds, which might otherwise peck at more vital parts of its body. The moth can then escape even if its wing is slightly torn. The main part of the wings shows a disruptive pattern when the insect is at rest.

Disruptive wing pattern

CITY MOTH, COUNTRY MOTH *right*

Some years ago, it was realized that the city form of the Peppered moth, *Biston betularius* (Europe), had gradually changed from a light to a black color. This was to escape birds, which could easily spot a light-colored moth on a smoke-polluted tree. In the countryside the same moth is still speckled white.

Black form

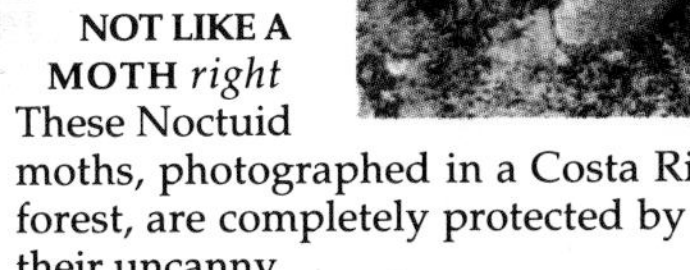

NOT LIKE A MOTH *right*

These Noctuid moths, photographed in a Costa Rican rain forest, are completely protected by their uncanny resemblance to lichen on lichen-covered bark.

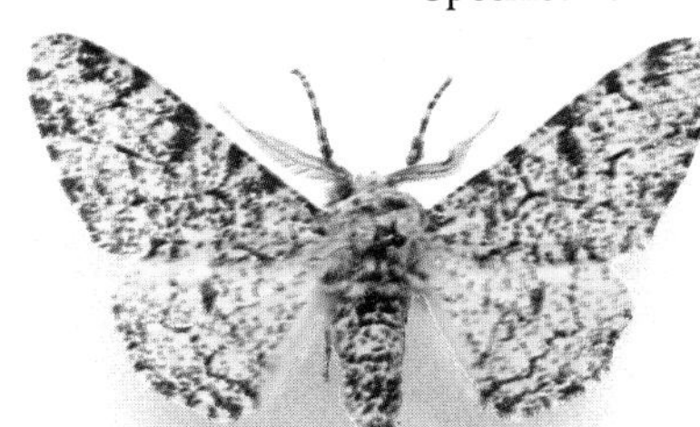

Speckled form

DISAPPEARING CATERPILLAR *left*

By blending with the bark of a tree, this Lappet moth caterpillar (species unknown) is completely protected from predators during daylight hours.

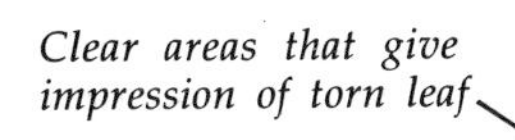

DECAYING LEAF *above*

This South American Leaf moth, *Belenoptera sanguine*, reproduces a "dead-leaf" pattern on its wings, including the "skeletonized" part often found on dead leaves. When resting, the moth rolls the front part of its wings to resemble a leaf stalk.

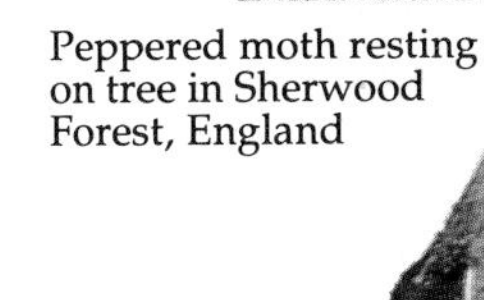

Peppered moth resting on tree in Sherwood Forest, England

Clear areas that give impression of torn leaf

DAMAGED LEAF *above*

To help it look more like a leaf, this green Pyralid moth, *Siga liris* (S. America), has irregularly shaped clear areas in its wings. When the moth is resting, these give the impression of a damaged leaf.

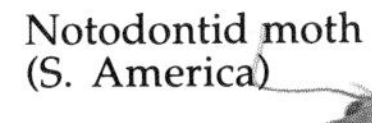

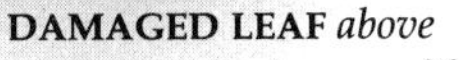

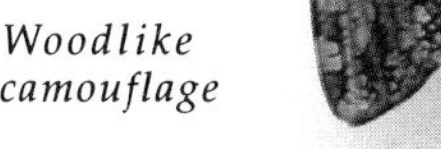

Notodontid moth (S. America)

Woodlike camouflage

Notodontid moth (S. America)

WOOD BORER *above*

The caterpillar of this Carpenter moth (C. America) bores into trees. As an adult, the moth is almost invisible against bark (right).

VANISHING MOTHS

These three pinned moths have been left with their wings in a normal resting position to show how successful their camouflage is. In order to survive, they must not look like moths or they would soon be detected by a hungry bird or lizard.

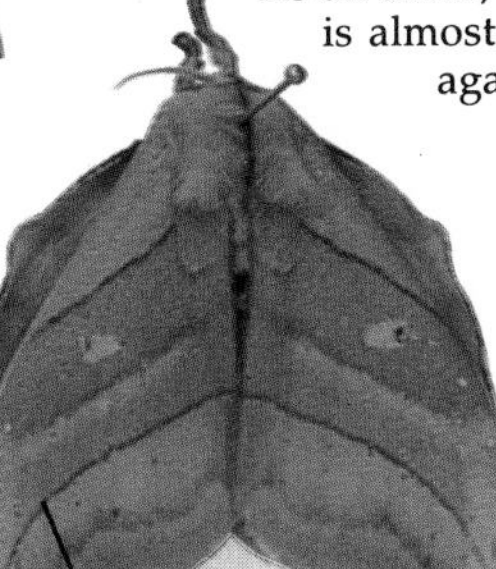

Dead-leaf-like camouflage

Saturniid moth, *Automeris* species (S. America)

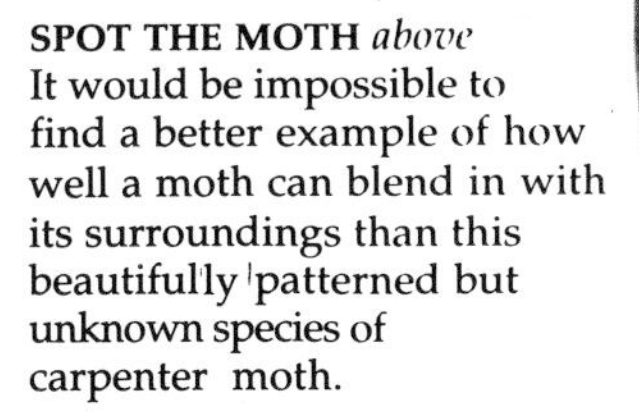

SPOT THE MOTH *above*

It would be impossible to find a better example of how well a moth can blend in with its surroundings than this beautifully patterned but unknown species of carpenter moth.

Mimicry and other unusual behavior

ALTHOUGH MOST BUTTERFLIES and moths live "normal" lives, there are species that behave in unusual ways. They include moths that swim underwater, and others whose caterpillars live in ants' nests or beehives. There is also the amazing way in which some species of Lepidoptera mimic other species. If some tropical butterflies seem to advertise themselves with their bright colors and slow flight, it is usually because they are poisonous to predators. But don't be fooled by the "same species" of butterfly flying nearby; it is not poisonous at all, just a very good mimic.

Fruit with hole where adult moth (above right) has pushed its way out after pupation

Caterpillar

JUMPING BEAN MOTH
Jumping Beans, or "vest-pocket pets," are not beans at all but the caterpillar of *Cydia saltitans* (C. America). The "bean" is a small fruit that the caterpillar has bored its way into. When placed near heat, the caterpillar twists and jumps. This may be to move the fruit, with the caterpillar inside, out of direct sunlight.

SUGAR CANE PEST
The caterpillars of this Galleriid moth, *Eldana saacharina*, are a menace in Africa because they bore into sugar cane stems.

DANGEROUS TO CATTLE
This Pyraustine moth, *Filodes sexpunctalis*, is a species of moth that uses its proboscis to feed on the liquid around the eyes of cattle. In doing so it transmits diseases.

FLOATING MOTH
The caterpillar of this Brown Chinamark moth, *Elophila nympheata* (Europe & Asia), lives in water plants (see opposite page).

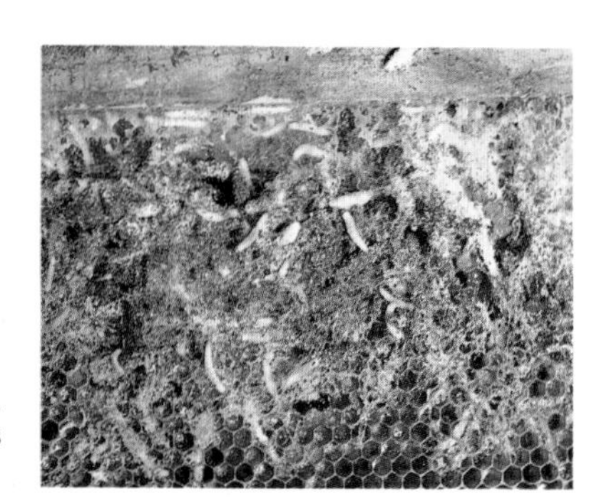

AN ENEMY IN THE HIVE
Beekeepers have to watch out for the caterpillars of the Wax moth, *Galleria* (N. America). The caterpillar not only feeds on the wax but destroys the honeycomb by making silk-lined galleries (right).

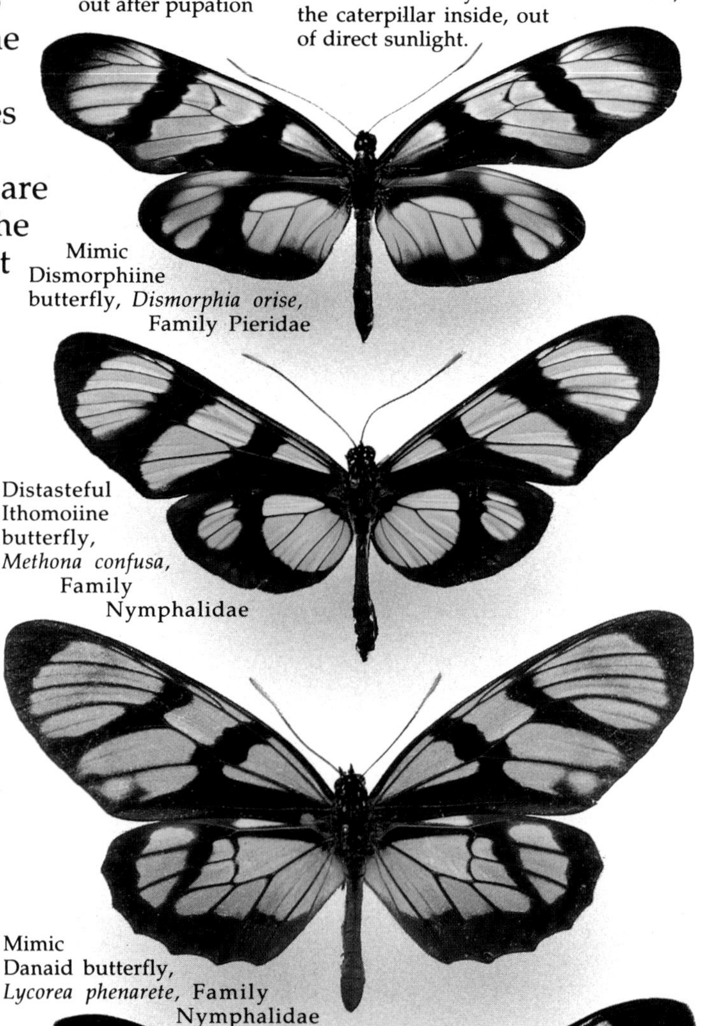

Mimic Dismorphiine butterfly, *Dismorphia orise*, Family Pieridae

Distasteful Ithomoiine butterfly, *Methona confusa*, Family Nymphalidae

Mimic Danaid butterfly, *Lycorea phenarete*, Family Nymphalidae

Mimic Castniid moth, *Gazera linus*, Family Castniidae

COPYING YOUR NEIGHBOR
Although these four South American "butterflies" look amazingly alike, not only do they all belong to different families, but one of them is in fact a moth. The moth and two of the butterflies are protected because the Ithomiine butterfly they mimic is distasteful to predators (see opposite page for more on mimicry).

Noctuid moth, (species and place of origin unknown)

Scent tuft has been artificially turned outward

Arctiid moth, *Creatonotos gangis* (Australia & Asia)

USING THE RIGHT SCENT
The strange-looking tufts protruding from both these moths' abdomens are known as hair pencils. Evident only in certain species of butterfly and moth, the tufts are a way for the male to disperse his scent and attract a female during courtship.

Transparent wings covered with fine hairs but few scales

Scales

GHOST OF THE RAIN FOREST

The Clearwing butterfly, *Haetera macleannania,* lives in the jungles of South and Central America, where its ghostlike appearance makes it almost invisible. Butterflies and moths with transparent wings have fewer scales than other Lepidoptera.

A WATER PLANT SHELTER

This is the adult Brown China-mark moth (see opposite page) and its caterpillar, after the larva has built a shelter in a water plant.

Caterpillar in leafy shelter on underside of water plant

SILVER FLASH

The Silver butterfly, *Argyrophorus argenteus,* is one of the most spectacular South American butterflies. As it flies in the high mountain areas of the Andes, its silvery metallic wings reflect the light, making the butterfly seem to appear and disappear.

SEASONAL SWITCH *below and below left*

Although the two butterflies on the left look very different, they are in fact both Pansy butterflies, *Precis octavia* (Africa). The Pansy has a different pattern if it develops in the wet rather than the dry season.

Dry-season form

Wing-coupling mechanism like that of moths

THE OLDEST AUSTRALIAN? *above*

Because it is the only butterfly with a wing mechanism like a moth's, the Australian Regent Skipper butterfly, *Euschemon rafflesia,* may be one of the world's most primitive butterflies.

Wet-season form

A BAD-TEMPERED BABY *below*

It may look sweet but the Saddleback caterpillar of this tropical Limacodid moth is best left alone; the hairs on its spines are poisonous to the touch.

PRIZE MIMICS *below*

In this mimicry chain the Ecuador Small Postman butterfly mimics its equally poisonous rain forest neighbor, the Ecuador Postman butterfly. The two Brazilian subspecies (right) look even more alike.

Mimicry

One of the most dramatic ways that certain species of butterflies and moths protect themselves is by mimicking neighboring species of Lepidoptera. This often takes the form of an edible species copying the color pattern of a species with an unpleasant smell or taste. Birds and other predators learn to recognize the warning patterns or bright colors of the harmful species and leave it alone - a form of protection that extends to the mimic species that is edible.

Small Postman butterfly, *Heliconius erato* (Ecuador)

Postman butterfly, *Heliconius melpomene* (Ecuador)

Small Postman butterfly, *Heliconius erato* (Brazil)

Postman butterfly, *Heliconius melpomene* (Brazil)

Endangered species

DEPENDENT AS THEY ARE ON WILD PLANTS and open countryside, butterflies and moths are extremely vulnerable to changes in the environment, especially those caused by human beings. In recent times many of these beautiful insects have become first rare, then endangered, then extinct. Butterflies and moths are more common in the tropics than in Europe or North America, but even there the destruction of the rain forests has reduced their numbers and variety. In milder climates people's need for land has meant that many more insects have become classified as endangered species. Many of these species are listed in the Red Data Book of the International Union for the Conservation of Nature.

BUTTERFLY FRAUD *above*
This famous butterfly fraud dates from about 1702. After painting eyespots on the wings of Brimstone butterflies, the "collector" claimed they were a new species of butterfly, later described as *"Papilio ecclipsis."*

Badly torn wing

Large wings and long tails typical of swallowtails

RECENT DISCOVERY
The Philippine Swallowtail butterfly, *Papilio chikae*, has only recently been discovered. Not only is its habitat threatened, but it is also in danger from keen collectors.

EARLY COLLECTOR *below*
Our knowledge of Lepidoptera is based largely on the work of early collectors such as the Englishman Moses Harris (left), whose classic book, *The Aurelian*, was published in 1766.

Old engraving of pinned butterfly

The only European Saturniid moth with tails on its hind wings

LONG-TAILED BEAUTY
Surely the most beautiful European moth, the Spanish Moon moth, *Graellsia isabellae* (French Alps & central Spain), now has to be protected by law.

Innocent casualties

"As dead as a dodo" is the sad phrase we use for a creature that has become extinct. In the United States the Xerces Blue butterfly has vanished; in Britain several once-common species are now extinct. In Britain enthusiasts have tried to introduce related subspecies from the European mainland to replace the vanished British species. But in other parts of the world extinct species cannot be replaced, and many beautiful insects are now just as dead as the dodo.

The Essex Emerald moth, *Thetidia smaragdaria* (Europe), is now very rare in Britain

Engraving of 19th-century collecting box (left)

Once found in coastal sand dunes of California

IN MEMORY OF
The Xerces Society, a worldwide conservation group, was formed in memory of the Californian Xerces Blue, *Glauscopsyche xerces*, last seen near San Francisco in 1941.

MAKING A COMEBACK *left*
After their habitats were largely destroyed, the Large Copper and the Large Blue became extinct in Britain. Now, subspecies of both butterflies have been introduced from the European mainland into specially chosen areas of England.

The Large Blue butterfly, *Maculinea arion* (Europe), was recently reintroduced into southwest England

The beautiful Large Copper butterfly, *Lycaena dispar* (Europe), became extinct in Britain in the 1800s (p. 28)

THE DISAPPEARING AMERICAN
The Regal Fritillary butterfly, *Speyeria idalia*, occurs in Canada and a number of American states. But as its natural grassland habitat is plowed up, the butterfly has become increasingly rare.

19TH-CENTURY COLLECTOR
The Muséum Nationale d'Histoire Naturelle, Paris; the Smithsonian, Washington, D.C; the Natural History Museum, London; and other great national collections were all built up by dedicated collectors.

NO FOREST, NO BUTTERFLIES *above*
A far greater threat to butterflies and moths than collectors or disease has been the increasing destruction of their habitats. Many countries now have laws preventing excessive collecting, and so fewer species have been destroyed by collectors. But throughout the world, important habitats, like this Central American rain forest, are being lost to new farmlands and towns. Many harmless insects are also killed by herbicides and insecticides.

MOUNTAIN RARITY *left*
The Corsican Swallowtail butterfly, *Papilio hospiton*, is found only in the mountainous regions of Corsica and Sardinia. Never numerous, its collection is now forbidden by law.

BEAUTIFUL BIRDWING
One of the largest known butterflies, the Queen Alexandra's Birdwing, *Ornithoptera alexandrae* (New Guinea), has suffered from the destruction of its forest habitat and the activities of collectors.

COLLECTORS' FAVORITE
Found only in Jamaica, the large Homerus Swallowtail butterfly, *Papilio homerus*, is unfortunately popular with collectors. It is now on the list of endangered butterfly species.

NO LONGER A PEST *below*
Increasingly scarce in mainland Europe, the Black-veined White butterfly, *Aporia crataegi*, is already extinct in Britain. Its caterpillars used to be a pest of fruit trees.

HERE TODAY, GONE TOMORROW?
Although the Zebra Swallowtail, *Eurytides marcellinus*, can be found in several parts of Canada and the United States, it is threatened by the destruction of its food plant, as well as by suburban growth.

Watching butterflies and moths

FOR MANY YEARS people have collected butterflies and moths as a hobby and for scientific study. But it is better for the insects themselves, as well as more interesting and rewarding, to watch them in the field. You can also photograph butterflies or catch them in a net for closer examination before releasing them again. Most moths fly at night, but it is possible to study members of day-flying families. When watching butterflies and moths, you can discover the answers to many questions about their behavior. Do they feed at a certain time of day? Do they have a territory, and if so, how do they defend it? Do they migrate, and if so, when? Do their flight patterns change in different seasons? Studying butterflies in this way is simple: you do not need much equipment, and you will not harm them - all you need is patience.

A 19th-century interpretation of the fashion for insect collecting

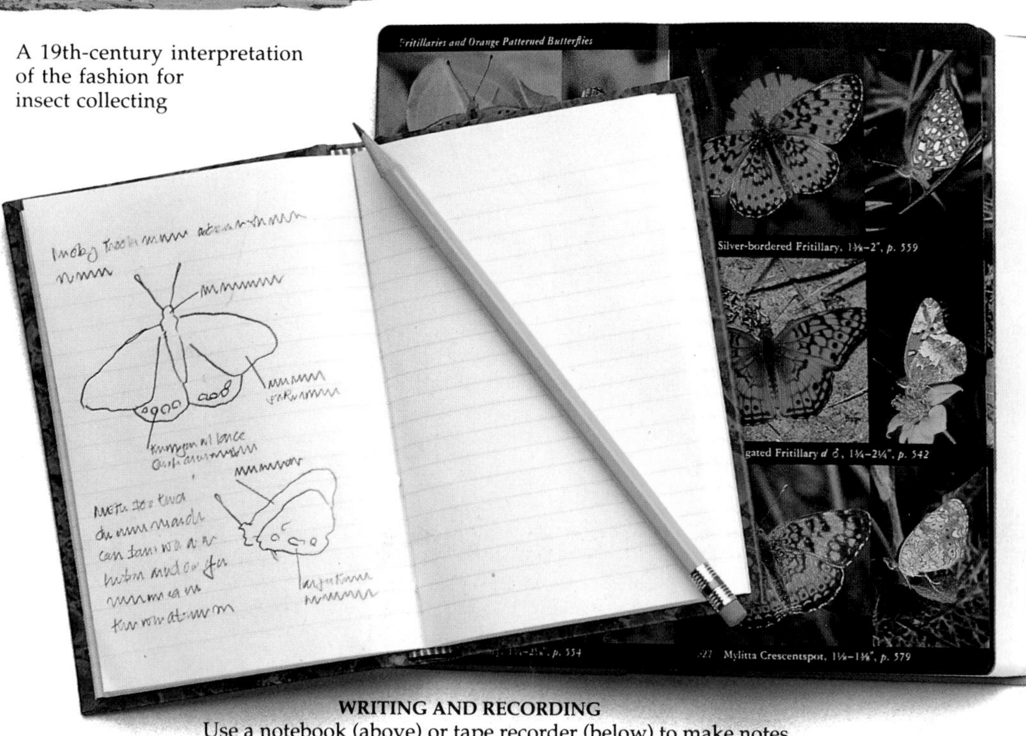

WRITING AND RECORDING
Use a notebook (above) or tape recorder (below) to make notes on each butterfly's appearance and behavior, together with the date, time, weather conditions, and details of the locality. A good field guide helps in identification.

PINNING THEM DOWN
Compiling a butterfly collection was once a popular pastime.

WARNING FOR COLLECTORS
Many butterflies are protected by law, and in some countries collecting is forbidden. Many places require a permit for the study of rare species. Study on nature reserves is usually also carefully controlled. A number of butterflies cannot be collected and sent out of the country without a permit. International conventions regulate trade in endangered species. Always check the local regulations.

Macro lens (which enlarges images) for close-ups

CLOSE-UP CAMERA
An SLR camera with a macro lens is ideal. Bright sunlight will give the best results; use flash in the shade.

SHORT-FOCUS TELESCOPE
In the field, use an 8 x 20 short-focus optical telescope (which has a magnification of x 8 and a lens diameter of 20 mm).

PILL BOXES
Glass or plastic lids allow you to examine an insect and later to release it unharmed. Do not keep specimens in pill boxes for too long.

Pill boxes

IN FULL PURSUIT
This 19th-century collector is using a large clap net, a type of net formerly used for catching birds. In the 1800s, people saw insect collecting as a harmless hobby.

Collecting jar

USING THE NET
Sweep the net through the air in the direction shown. As you double the net back, the butterfly will be directed toward the bottom. A final rapid flick of the wrist traps the insect in the net. When using the net in this way, be careful to avoid thorns and sharp twigs that may damage both the mesh and the insect.

COLLECTING JARS
Insert a twig for the insect to perch on in each jar. The specimen will then keep still and not damage itself by fluttering about.

Rounded end of bag reduces risk of damaging butterfly

CLOSE-UP VIEW
Using a magnifying glass is often the only way to see specialized features mentioned in books.

Long bag made of fine material to protect butterfly; dark mesh is less noticeable than lighter fabric

General-purpose sweep net for collecting insects - not suitable for butterflies

Emperor net

FOR HIGH-FLYERS
The emperor-net was devised to catch butterflies that lived in the treetops, such as the Purple Emperor, *Apatura iris* (Europe & Asia; p. 29).

Rearing butterflies and moths

An elaborate 19th-century "caterpillar house"

MANY BUTTERFLIES AND MOTHS are easy to rear from egg to adult, provided that the basic conditions are right. You should always handle the insects with care, keep the temperature close to that of the natural habitat, and give each species its own specific food plant. In addition, food for caterpillars must be fresh - either in the form of freshly plucked leaves or complete, pot-grown plants. Sometimes the food may appear to be in good condition but the caterpillars will not eat it. The reason is usually the state of the plant - if it is short of water, or if it is too old, the caterpillars may reject it. We still know little about the exact needs of plant-feeding caterpillars: although some are choosy about their food, other species, such as some of the Pyralid and Tineid moths, whose caterpillars feed on grain or flour, are easy to rear. Adult butterflies and moths, if they take any food at all, will sometimes feed from cut flowers or the flowers of potted plants. Many will sip happily at a weak solution of sugar or honey in water. But in general you should not keep adults in captivity for long. Once they have emerged, release them when weather conditions are right.

Keeping caterpillars

Keep caterpillars in a special cage or in a muslin sleeve over the food plant itself. Although many species take food from only one plant, some caterpillars will eat a wide variety of food. Given the opportunity, certain caterpillars will even eat objects they cannot survive on, such as plastics and man-made fibers.

FINE MESH CAGE
The soft walls of a fine mesh caterpillar cage protect its delicate inhabitants. A zip fastener gives you access to the contents; paper catches any debris.

Food plant

Newspaper

Zipper opening

HOMEGROWN FOOD
Foods grown in the cage itself will be attractive to most caterpillars, but the larvae eat so much that you should keep extra plants in reserve.

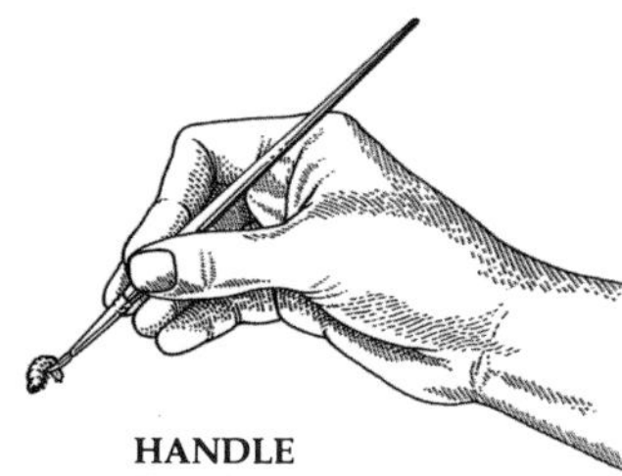

HANDLE WITH CARE
Caterpillars are very delicate. The safest way to pick them up is with a fine paintbrush. Some larvae have stinging hairs - another reason for handling them in this way.

PLANT-POT CAGE
You can attach a sleeve of muslin to a small pot and rear caterpillars on a growing plant. Take care that the inmates do not strip all the leaves and kill the plant.

Keeping adults

When the adult insects have emerged from their pupae, they will require a different food source. Either a selection of cut flowers or a weak solution of honey and water can be used. Alternatively, you can provide fruit juice (see below). Butterflies may mate before you release them, providing eggs so that you can start the life cycle over again.

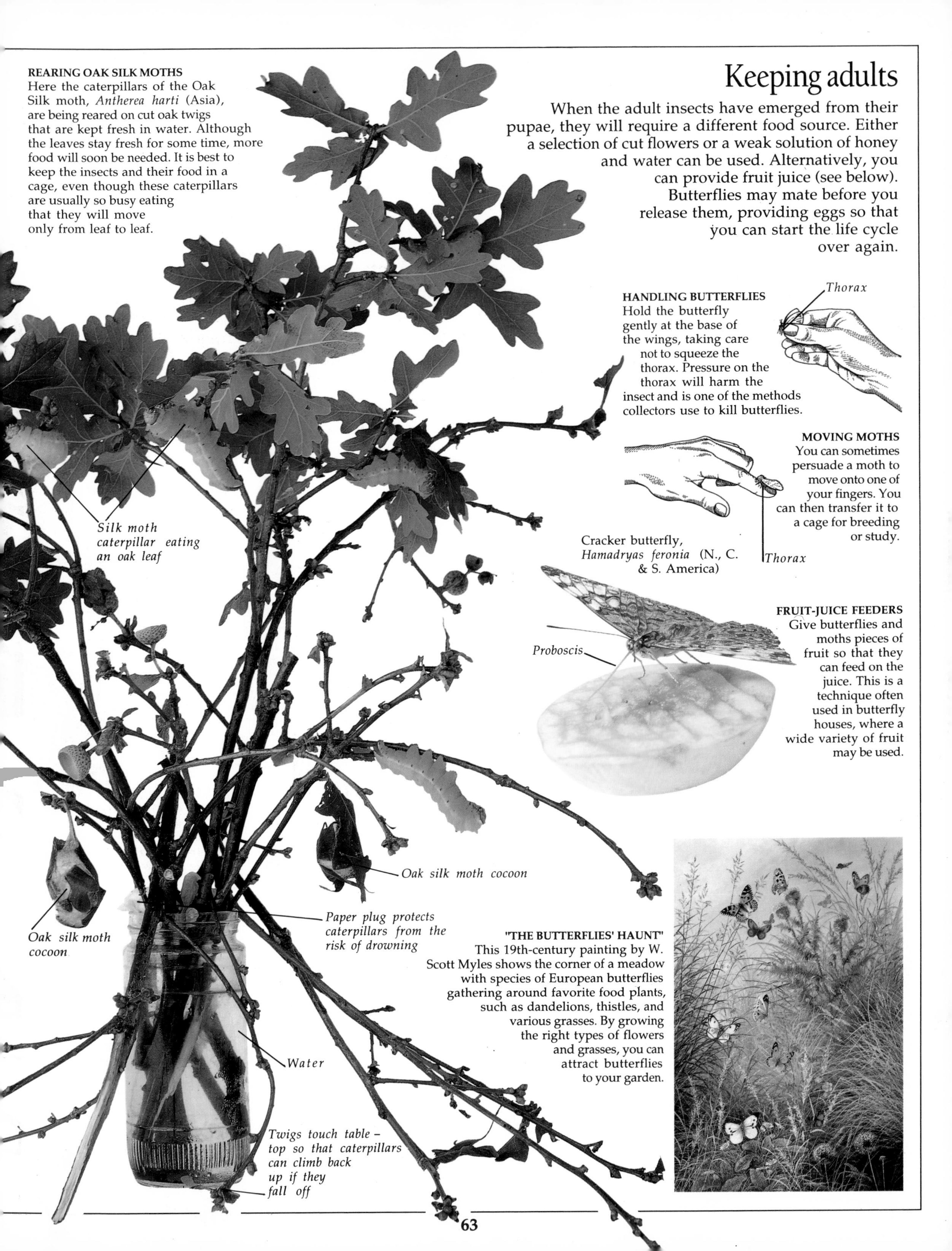

REARING OAK SILK MOTHS
Here the caterpillars of the Oak Silk moth, *Antherea harti* (Asia), are being reared on cut oak twigs that are kept fresh in water. Although the leaves stay fresh for some time, more food will soon be needed. It is best to keep the insects and their food in a cage, even though these caterpillars are usually so busy eating that they will move only from leaf to leaf.

HANDLING BUTTERFLIES
Hold the butterfly gently at the base of the wings, taking care not to squeeze the thorax. Pressure on the thorax will harm the insect and is one of the methods collectors use to kill butterflies.

MOVING MOTHS
You can sometimes persuade a moth to move onto one of your fingers. You can then transfer it to a cage for breeding or study.

Cracker butterfly, *Hamadryas feronia* (N., C. & S. America)

FRUIT-JUICE FEEDERS
Give butterflies and moths pieces of fruit so that they can feed on the juice. This is a technique often used in butterfly houses, where a wide variety of fruit may be used.

"THE BUTTERFLIES' HAUNT"
This 19th-century painting by W. Scott Myles shows the corner of a meadow with species of European butterflies gathering around favorite food plants, such as dandelions, thistles, and various grasses. By growing the right types of flowers and grasses, you can attract butterflies to your garden.

Index

Acknowledgments

Dorling Kindersley would like to thank:
The staff at the London Butterfly House; also Tom Fox for his advice, and David Lees for his advice and for providing butterfly eggs and caterpillars.
The staff at the British Museum (Natural History), in particular Mr. P. Ackery, Mr D. Carter, Miss J. Goode, Mr C. Owen, Mr. C. Smith and Mr A. Watson.
Colin Mays of Worldwide Butterflies for providing the silkworms on pp. 40-41.
Stephen Bull, and Fred Ford and Mike Pilley of Radius Graphics for artwork.

Picture credits
t=top b=bottom m=middle l=left r=right c=centre

Ardea London: 26br; 27br; 35tc; 37mr; 50ml; 51mr, br; 52mr; 55ml, bl; 57mr
The Bridgeman Art Library: 6tl; 24t
The Trustees of the British Museum: 6br
The British Museum (Natural History): 58t, mc
Professor Frank Carpenter: 6bl
E.T. Archives: 12t; 49mr
Fine Art Photographs: 63br
Heather Angel: 26bl; 27bl
Jeremy Thomas/Biofotos: 31tr
John Freeman London: 34ml; 45tr
Mansell Collection: 14t; 40tl, bl
Mary Evans Picture Library: 10t; 40tr; 56tl; 60tl
Oxford Scientific Films Ltd: 11bl; 45mr; 55tc, mr; 56mc; 59tl
Paul Whalley: 29tc; 38ml; 59bc; 61tr
Quadrant Picture Library: 44ml
Royal Botanic Gardens, Kew: 33mc
Sonia Halliday Photographs: 40m

Illustrations by: Coral Mula for pp. 61, 62-63; Sandra Pond for pp 50-51; Christine Robins for pp. 16-17, 18-19, 28-29, 30, 32.

Picture research by: Millie Trowbridge.

SS COLMAN-JOHN NEUMANN
T 3510